Assessing Building Performance

Assessing Building Performance

Edited by

Wolfgang F.E. Preiser
Jacqueline C. Vischer

ELSEVIER
BUTTERWORTH
HEINEMANN

AMSTERDAM   BOSTON   HEIDELBERG   LONDON   NEW YORK   OXFORD
PARIS   SAN DIEGO   SAN FRANCISCO   SINGAPORE   SYDNEY   TOKYO

Elsevier Butterworth-Heinemann
Linacre House, Jordan Hill, Oxford OX2 8DP
30 Corporate Drive, Burlington, MA 01803

First published 2005

British Library Cataloguing in Publication Data
A catalogue record for this book is available from the British Library

ISBN 0 7506 6174 7

For information on all Elsevier Butterworth-Heinemann publications
visit our website at http://books.elsevier.com

Typeset by Charon Tec Pvt. Ltd, Chennai, India
www.charontec.com

Front cover: the Aronoff Center for Design and Art at the University of Cincinnati.
Building designed by Peter Eisenman

Working together to grow
libraries in developing countries

www.elsevier.com | www.bookaid.org | www.sabre.org

ELSEVIER   BOOK AID
            International   Sabre Foundation

Contents

Foreword

Why have architects talked about the assessment of building performance for so long and yet have been so slow to do anything about it? This question is particularly acute in office design, a topic that many of the chapters in this book address. Conventional office design and space planning are being challenged more and more by the new ways of working that ubiquitous information technology is making attractive and accessible to many clients and users. Old rules of thumb may not be working so well these days but they linger on in many design practices.

Post-occupancy evaluation is certainly considered by many designers and clients to be too time-consuming and expensive. Simply facing up to the reality of having to change may in itself be enough of an obstacle in the lives of busy professionals. Putting oneself in the position of potentially having to admit errors and thus opening the way to blame or even litigation may be a fear that is not even easy to admit.

These are real considerations but there are three deeper explanations all of which are addressed in this excellent book. The first is that both organizations and buildings are highly complex phenomena, not least because they are saturated by values and motives. The changing relationship between them over time makes them even harder to study and explain. Consequently and inevitably assessing building performance pushes the frontiers of social science. The second reason is that architects and designers, and many clients too, suffer from what might be called the curse of the project. Because of the ways in which design professionals, facilities managers and corporate real estate executives are constrained to work, it becomes almost impossible, operationally, day by day, for them to conceive of life as anything more than an unending series of separate, sequentially experienced projects. Generalisations become very hard to make. This quasi psychological, semi pathological condition is aggravated by the third and most fundamental reason for the general failure, so far, to put building performance assessment into common practice: the chronically fragmented and confrontational nature of the construction industry itself and, even worse, of its relationship with its clients. Supply side behaviour has become endemic.

All of which paradoxically makes me very optimistic. It may have taken a long time but the fact that these issues are faced up to so intelligently and vigorously in this collection of chapters must surely mean that things are now on the move. The structure of the book demonstrates a firm grip on the realities of the design and building process and is in itself a manifesto for change. The chapters reveal a universal sensitivity to the demand side of emerging user concerns and priorities. Epistemological difficulties are faced up to honestly. The case studies are a model of how to communicate complex data about real life situations in which supply and demand considerations are woven together. The international provenance of the authors ensures the diversity and relativity that are so essential given the challenging task of assessing building performance.

This book turns difficulties into opportunities. It makes one almost proud to be an architect.

Francis Duffy
4 June 2004

Francis Duffy is a founder of DEGW, an international architectural practice that concentrates on the design of working and learning environments and carries out user research and brief writing as well as interior design and architecture. Duffy has been President of the Royal Institute of British Architects (RIBA) and of the Architects' Council of Europe. He has recently returned to DEGW London after a three year secondment to DEGW North America's office in New York. He is a Visiting Professor at MIT.

Preface

Convergence may be the proper term for describing how this book originated, both in terms of the contributing authors, and the timing. Both co-editors were actively involved in the early days of post-occupancy evaluation in the late 1960s, but came to this sub-discipline from different backgrounds: Wolfgang Preiser from architecture, and Jacqueline Vischer from environmental psychology. Over the course of the decades, both co-editors developed evaluation methodologies, both of which are now well accepted and in use around the world. In fact, the post-occupancy evaluation methodology developed by Preiser is now part of the professional development monograph series of the National Council of Architectural Registration Boards (NCARB). This means that a journey that started after graduating from Virginia Tech with a Masters of Architecture in environmental systems in 1969 continued for over 35 years, studying, writing about and pursuing the topics of post-occupancy and building performance evaluation. It has culminated at a point where every architect can learn about this subject and be tested on it for continuing education credit.

Jacqueline Vischer has been on a similar journey. Starting out with her doctoral research, focused on people's use of space in community mental health centres and psychiatric settings, she has developed a career and a mission around the human aspects of the built environment. Having studied, written about and pursued residential environments, prison architecture and hospital design, she has been engaged for the last ten years in user evaluation of the work environment, and has published three books on that subject, with a fourth due out soon.

Over the years, the co-editors have produced a number of collaborative efforts, including chapters in and the epilogue for the book *Building Evaluation* (Preiser, 1989), and the book *Design Intervention: Toward a More Humane Architecture* (Preiser and Vischer, 1991), which contains the precursor for the Building Performance Evaluation (BPE) conceptual framework presented in this book.

Then, in 1995, Preiser founded the International Building Performance Evaluation (IBPE) consortium, which sponsored symposia at international research conferences, such as those of the Environmental Design Research Association (EDRA) and the International Association for People-Environment Studies (IAPS). Most authors in this book, including the co-editor, joined this consortium and have contributed both methodological and case study materials over the years, which then became the foundation of this book.

The team of authors convened to contribute to this volume all have distinguished credentials. They come from a mixture of academic and practitioner backgrounds, with the vast majority being qualified architects. Experienced in POE activities and the BPE approach, they have successfully applied these to their work in Germany, the UK, the USA, Canada, Hong Kong, Brazil, Israel, Japan, and the Netherlands. They are all, in their own ways, pioneers and visionaries, who are holding out for a better quality and more humanistic buildings in the future. The chapters they have contributed reflect years of experience and a thorough working knowledge of their professions, and of the context within which they teach and work. In addition, the basic concepts of BPE are common to all.

This book, while entirely new in its conception and focused on a systematic approach to building performance evaluation, can be seen as a sequel to the book *Building Evaluation* (Preiser, 1989), and as a companion volume to *Improving Building Performance* (NCARB, 2003). With its global perspective and contributions from industrialized and industrializing countries, this book hopes to make a significant contribution by encouraging international collaboration in the quest for continuous quality improvement in the built environment.

Cincinnati and Montreal
March 2004

List of contributors

Cláudia Miranda de Andrade has a degree in Architecture and Urbanism from the Bennett Methodist University in Rio de Janeiro, a Masters degree in Architecture at the School of Architecture and Urbanism of the University of São Paulo, Brazil, and is currently a doctoral candidate at the same university. She is a partner and director of Saturno Planejamento, Arquitetura e Consultoria, a São-Paulo based firm specializing in performance evaluation, planning and interior design of office workplace environments. claudia.andrade@uol.com.br

Dr Bill Bordass runs William Bordass Associates, which undertakes field, desk and case studies of building performance; helps to manage and interpret research; and assists clients with briefing/programming and reviewing projects. bilbordass@aol.com

Adrian Leaman, BA, FRGS, FRSA, runs Building Use Studies, which carries out studies on buildings primarily from the users' point of view. al@usablebuildings.co.uk

Bill Bordass and Adrian Leaman have worked together on many projects, including Probe and the Feedback Portfolio described in Chapter 7.

Stephen Bradley is an architect specializing in workplace occupancy issues from a design and management perspective. Stephen has ten years of international experience in commercial architectural and engineering practice. He graduated in architecture from Bristol University, with further education at INSEAD business school in France. He holds a Master of Corporate Real Estate designation. Stephen teaches and writes frequently on workplace issues and is involved in research with several academic and industry groups, including editorial contribution to the British Council for Offices (BCO) guide to good practice in office specification. mail@aleximarmot.com

Gerald Davis, F-ASTM, F-IFMA, CFM, AIA, is President/CEO of the International Centre for Facilities. Davis is Chair of the ASTM Committee E06.25 on Whole Buildings and Facilities, current USA voting delegate of the ISO Technical Committee 59 on Building Construction, and of the Subcommittee 3 on Functional/User Requirements and Performance in Building Construction. He is the recipient of the Environmental Design Research (EDRA) Lifetime Achievement Award 1997; IFMA Chairman's Citation 1998; ASTM Fellow 1995; and IFMA Fellow 1999. info@icf-cebe.com

Fehmi Dogan recently received his PhD in Architecture from the Georgia Institute of Technology and is now assistant professor at the Ismir Institute of Technology, Ismir, Turkey.

Anne van Dortmont is a senior consultant at Van Wagenberg Associates, a consultancy practice for strategic facility management. She is active in the areas of facility organizing, master planning, design/redesign of (innovative, healthy and sustainable) offices and hospitals, building programming and evaluation of facilities. Her evaluation research focuses on the strategic building performance of work environments (i.e. innovative measurements and office concepts) with respect to organizational goals and strategy, as well as to individuals. anne@wagenberg.nl

Dennis Dunne is a consultant and was until recently the Chief Deputy of the California Department of General Services. He introduced BPE in the California Department of Corrections, Santa Clara County and the Department of General Services.

Joanna Eley is an architect with over twenty years' experience of advising organizations on strategies for space and facilities management. Clients include large public and private sector organizations, as well as smaller groups and charitable foundations. Joanna graduated in PPE at Oxford before progressing to architecture at the University of Pennsylvania and London University. She is co-author with Alexi Marmot of *Understanding Offices: What every manager needs to know about office buildings* (London, Penguin Books, 1995) and *Office Space Planning: Designing for Tomorrow's Workplace* (New York, McGraw-Hill, 2000). mail@aleximarmot.com

Cheryl Fuller is a facilities planner, programmer and evaluator with Fuller, Coe & Associates, Sacramento, CA and has conducted BPE for several state and county agencies since 1986.

David M. Hammond is a senior programme manager in the office of civil engineering's *Shore Facilities Capital Asset Management Strategic Initiative*. He earned a Bachelor of Science degree from Cornell University and a Master of Science degree from Syracuse University in Information Resource Management. He is a graduate of the Information Resource Management College at the National Defense University and holds a Department of Defense Chief Information Officer Certificate. He is a member of the Federal Facilities Council's Emerging Technologies Committee and of the IAI Board of Directors (International Alliance for Interoperability) – North America.

Michael J. Holtz is president of Architectural Energy Corporation, an energy and environmental research and consulting company with offices in Boulder, Colorado; San Francisco, California; and Chicago, Illinois. Mr Holtz has been engaged in building performance evaluation, sustainable design assistance, commissioning, and energy research since 1973. He has held senior research and management positions with the AIA Research Corporation, the Solar Energy Research Institute and, since 1982, with the Architectural Energy Corporation. He holds a Bachelors and Masters of Architecture from Ball State University and the State University of New York, respectively. mholtz@archenergy.com

Kevin Kampschroer is Director of Research for the Public Buildings Service, US General Services Administration. Among other responsibilities he helped develop the innovative Workplace 20.20 Program that includes BPE as part of project delivery.

Dr Akikazu Kato, Department of Architecture and Civil Engineering, Toyohashi University of Technology, Japan, is involved in various research, planning and design projects pertaining to healthcare and workplace environments, as well as public housing. Currently, he is working towards development of complete course materials on Facility Management for both undergraduate and graduate students in Architecture and Civil Engineering. kato-a@acserv.tutrp.tut.ac.jp

Alex Lam is president and founder of the OCB Network in Toronto, Canada, a global consultancy specializing in business transformation and emotional intelligence in facility management. He received his Bachelor of Architecture degree from McGill University (1967) and a Master of Theological Studies from Ontario Theological Seminary (1995). He is a Certified EQ-*i* Administrator with Multi-Health Systems Inc. (MHS) in Toronto (2000) on emotional intelligence in workplace performance. A frequent speaker at international conferences, he has taught at the University of Hong Kong, the Polytechnic University of Hong Kong, the University of Manitoba and Ryerson Polytechnic University. alexatocb@aol.com

Dr Brenda C. Leite has a degree in Civil Engineering at the Federal University of Minas Gerais, Brazil. She is a professor at the Civil Construction Engineering Department, Polytechnical School of the University of São Paulo (EPUSP), Brazil. She has a Masters degree in Architecture from the School of Architecture and Urbanism at the University of São Paulo, Brazil and is a doctor in Mechanical Engineering at EPUSP, Brazil. bcleite@usp.br

Pieter C. Le Roux, Department of Architecture and Civil Engineering, Toyohashi University of Technology, Japan, is involved in performance evaluative research pertaining to workplace environments. His research interests include the study and evaluation of workplace environmental factors and standards, as well as the impact thereof on employee health, comfort, well-being, and ultimately, performance and productivity. He is currently working toward a PhD in the related field of Facilities Management and workplace planning methodology and evaluation. Pieter@acserv.tutrp.tut.ac.jp

Dr Shauna Mallory-Hill is an assistant professor at the University of Manitoba in Canada, and a doctoral fellow at the Eindhoven University of Technology in the Netherlands. Her main interest of research is on strategic building performance evaluation and the feeding-forward of knowledge into design practice and education. Coming from a background of developing and disseminating universal design codes and guidelines, Mallory-Hill has always been interested in the study of a wide range of human-built environment relationships. Her most recent investigations examine organizations, workers and innovative indoor environments. s_mallory-hill@umanitoba.ca

Dr Alexi Marmot established AMA Alexi Marmot Associates in 1990 to help organizations to make best use of their buildings by applying evidence-based design. She was educated as an architect and town planner in Sydney and California, where she also received her doctorate, before moving to London. She has taught at the Bartlett School of Architecture and Planning, University College London and was a director of Research and Development at DEGW. She serves on RIBA committees, has served on the Urban Design Group committee, and was appointed as a CABE enabler in 2001. Alexi is a frequent conference speaker and has prepared many articles on space design and management for professional journals, and has co-authored two books. mail@aleximarmot.com

Dr Sheila Walbe Ornstein has a degree in Architecture and Urbanism, as well as Masters and Doctorate degrees. She is also a professor at the School of Architecture and Urbanism at the University of São Paulo (FAUUSP), Brazil. She was the head of the Architectural Technology Department (1995–2002) and the Vice Dean (1998–2002) at FAUUSP. She is also a senior researcher and coordinator of the Post-Occupancy Evaluation (POE) programme at the Research Centre on Architecture and Urban Design Technology of the University of São Paulo (NUTAU–USP). Her research and teaching interests also include the study of construction technologies. sheilawo@usp.br

Dr Wolfgang F.E. Preiser is an educator, researcher and consultant, with a focus on Building Performance Evaluation/Post-Occupancy Evaluation and universal design. He has lectured at over 70 universities and institutions all over the world, and is widely published with 14 books to his credit and numerous articles in professional journals and conference proceedings including *Improving Building Performance* (2003), *Universal Design Handbook* (2001) and *Post-Occupancy Evaluation* (1988), which has been translated into Korean, Japanese, and Arabic. For the past 37 years, he has taught at Virginia Tech, Penn State, the University of Illinois, the University of New Mexico and, since 1990, the University of Cincinnati where he is a Professor of Architecture. wolfgang.preiser@uc.edu

Professor Susan Roaf was born in Penang, Malaysia, educated in Australia and the UK, and is co-director of the Oxford Centre for Sustainable Development: Architecture, at Oxford Brookes University. During ten years in Iran and Iraq she studied aspects of traditional technologies. Her current research interests include building sustainability, thermal comfort, photovoltaics, eco-tourism, water conservation, climate change, natural ventilation and passive buildings. She is an Oxford City Councillor and has recently published *Ecohouse Design* and *Benchmarking Sustainability*. sroaf@brookes.ac.uk

Dr Ulrich Schramm is Professor in the Department of Architecture and Civil Engineering at the University of Applied Sciences in Bielefeld, Germany. Previous positions include Assistant Professor at the University of Stuttgart and POE/programming specialist with Henn Architects in Munich. He received his PhD in Architecture from the University of Stuttgart and a post-doctoral fellowship from the German Research Foundation (Deutsche Forschungsgemeinschaft, DFG) for his stay at the University of Cincinnati as Visiting Professor of Architecture. ulrich-schramm@t-online.de

Françoise Szigeti is Vice President of the International Centre for Facilities. Szigeti is Vice Chair of the ASTM Subcommittee E06.25 on Whole Buildings and Facilities; and former Chair of the ASTM E06.94 on Terminology and Editorial. She is a co-author of the 1987 IFMA Benchmark Report, which was the first of its kind and was based on her proposal to IFMA, and of the Serviceability Tools & Methods (ST&M). She is the recipient of the Environmental Design Research (EDRA) Lifetime Achievement Award 1997.

Dr Danny S.S. Then is the Programme Leader of the Graduate Programme in Facility Management at the Hong Kong Polytechnic University. His experience in Facility Management educational development at postgraduate level includes Heriot-Watt University, Scotland (1985–96) and Queensland University of Technology, Australia (1996–2001). He is Joint Coordinator of CIB W070, Working Commission on Facilities Management and Asset Maintenance of the International Council for Research and Innovations in Building and Construction. He has published widely and is the co-author of *Facilities Management and the Business of Space* (Butterworth-Heinemann, 2001).

Kazuhisa Tsunekawa, Department of Architecture, Nagoya University, Japan, is a lecturer in Architecture and is also involved in various research, planning and design projects in workplace environments. He is currently in charge of campus master planning at Nagoya University, and is making efforts to realize Facility Management theory in practice. j45993a@cc.nagoya-u.ac.jp

Dr Jacqueline C. Vischer is an environmental psychologist with international research and consulting experience. She has published four books, three of which deal with design and use of environments for work, as well as numerous articles on designing new workspace, office building evaluation, users' needs in buildings, indoor air quality, user-manager communication, facilities management, and architectural programming. Since 1998, Dr Vischer has been a full-time Professor and Director of a new Interior Design programme at the University of Montreal. jacqueline.vischer@umontreal.ca and biubeu@wn.net

Dr Theo J.M. van der Voordt is associate professor and a senior research associate at the Faculty of Architecture of the Delft University of Technology. His research interests focus on briefing and post-occupancy evaluation. He developed design guidelines for healthcare centres, childcare centres, facilities for mentally retarded people, housing and care facilities for the elderly, universal access and crime prevention through environmental design. He is involved in research on workplace innovation, new office concepts and workplace performance indicators. d.j.m.vandervoordt@bk.tudelft.nl

Dr Rotraut Walden's major fields of research are Architectural Psychology, as well as Work and Organizational Psychology. She is Senior Lecturer for the Institute for Psychology at the University in Koblenz, Germany, where she holds tenure, and has been a member of the Environmental Design Research Association (EDRA) since 1989. She is the author of *Lively Dwelling: Development of Psychological Guidelines for Housing Quality,* as well as *Happiness and Unhappiness. Experiences of Happiness and Unhappiness from the Interactionistic Perspective,* and the co-author of *Psychology and the Built Environment, Places for Children* and *Schools of the Future.* walden@uni-koblenz.de

Dr Ahuva Windsor is an Environmental Psychologist based in Tel Aviv, Israel. For the past 16 years she has been practicing as an independent consultant in London, UK, and, since 1992, in Tel Aviv, Israel. She has been working on office buildings, as well as healthcare, educational and welfare facilities, conducting surveys and evaluations, and producing design guidelines. She also teaches at the School of Business Administration of College of Management in Rishon Lezion, Israel. ahuvaw@isdn.net.il

Dr Craig Zimring is an Environmental Psychologist and Professor of Architecture in the Georgia Institute of Technology, College of Architecture. craig.zimring@arch.gatech.edu

Acknowledgements

First of all, the editors would like to thank the contributors to this book for their positive and collaborative spirit in completing this project. This book is the result of a fast-track effort on the part of everyone involved: contributing authors, the editors, and Elsevier. The entire project took nine months before going into production, something that would not have been possible without speedy, electronic communications around the world – and this despite occasional computer glitches and meltdowns. An editorial system was devised, which allowed us to track the progress of each chapter's content and format through three-plus rounds of editing. We would like to thank our editor at Elsevier, Sarah Hunt, for her support throughout, and our project assistant, Dede Price, for helping prepare the manuscript in a professional and timely manner. Jay Yocis of the University of Cincinnati created the cover image. We also express heartfelt gratitude to the members of our respective families, who watched us struggle through yet another book project, and continued to offer advice, support and solace at opportune moments.

PART ONE

Introduction and Overview

1

The evolution of building performance evaluation: an introduction

Wolfgang F.E. Preiser and Jacqueline C. Vischer

Editorial comment

Building performance evaluation (BPE) is an innovative approach to the planning, design, construction and occupancy of buildings. It is based on feedback and evaluation at every phase of building delivery, ranging from strategic planning to occupancy, through the building's life cycle. It covers the useful life of a building from move-in to adaptive reuse or recycling. BPE came into being as a result of knowledge accumulating from years of post-occupancy studies of buildings, the results of which contained important information for architects, builders and others involved in the process of creating buildings – information that is infrequently accessed and rarely applied in most building projects. How then to systematize not only the research activities needed to acquire feedback from users at every stage, but also to ensure that such feedback is directly applied to the building delivery process, such that it is incorporated throughout?

BPE is a way of systematically ensuring that feedback is applied throughout the process, so that building quality is protected during planning and construction and, later, during occupation and operations. In this chapter, BPE is described, and reasons are given for why the building industry should make more use of this approach. The chapter traces the history and evolution of BPE from post-occupancy research, and outlines some of the methods available to the performance evaluation approach, many of which are illustrated in the chapters that follow.

1.1 Introduction

A rational building design process using feedback from ongoing evaluation can be conceptualized as a loop, whereby information fed back through continuous evaluation leads to better informed design assumptions, and ultimately, to better solutions. By using such a process, decision-makers are able to make better and more informed user-oriented design decisions. They are able to access building type-specific information gathered from evaluative research that is stored and updated in databases.

Different theoretical approaches to BPE were first presented in the book *Building Evaluation* (Preiser, 1989). Since then, there is not only increased interest and activity in this area of concern, both in the private and public sectors – for example, *Learning From our Buildings* (Federal Facilities Council, 2001) – but post-occupancy evaluation also continues to expand in the United States and other, mostly industrialized, nations around the world. Examples from four continents are included in this book. In addition, the National Council of Architectural Registration Boards (NCARB, 2003) has published a book on *Improving Building Performance*, which allows every architect in the USA to study and be tested in this field of endeavour, as part of professional development and continuing education.

The theoretical foundation of BPE is adapted from the interdisciplinary field of cybernetics, which is defined as 'the study of human control functions and of mechanical and electronic systems designed to replace them, involving the application of statistical mechanics to communication engineering' (Infoplease Dictionary, 2003). A systems model is proposed that is appropriate for this field because it holistically links diverse phenomena that influence relationships between people, processes and their surroundings, including the physical, social and cultural environments. Like any other living species, humans are organisms adjusting to a dynamic, ever-changing environment, and the interactive nature of relationships between people and their surroundings is usefully represented by the systems concept. Specifically, the systems approach to environmental research studies the impact of human actions on the physical environment – both built and natural – and vice versa. BPE has built on this tradition.

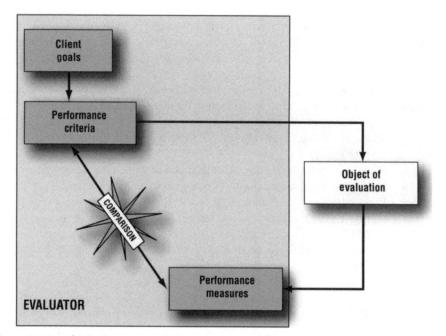

Figure 1.1 Basic feedback system.
Source: Architectural Research Consultants, Albuquerque, NM.

It is multi-disciplinary and it generates mostly applied research that, until recently, lacked a coherent theoretical framework.

The nature of basic feedback systems was discussed by von Foerster (1985). In the context of the building industry (Preiser, 1991, 2001), the strategic planner, programmer, designer, or other process leader is the effector or driver of the system (see also Figure 9.4 in Chapter 9). In the context of BPE, this can be personnel responsible for any or all phases of building delivery, including the evaluator, who makes comparisons between outcomes which are sensed or experienced by users, and the project goals expressed as performance criteria. In the case of building design, goals and performance criteria are usually documented in the functional programme or brief, and made explicit through performance language, as opposed to specifications for particular solutions and hardware systems, the selection of which are the domain of the designer.

1.2 Performance levels: a hierarchy of users' needs and priorities

The human needs that arise out of users' interactions with a range of settings in the built environment are redefined as performance levels. Grossly analogous to the human needs hierarchy (Maslow, 1948) of self-actualization, love, esteem, safety, and physiological needs, a tripartite breakdown of users' needs and their respective performance criteria parallels the three categories of criteria for evaluating building quality postulated centuries ago by the Roman architect Vitruvius (1960). These are firmness, commodity, and delight. This historic approach to setting priorities on building performance has been transformed into a hierarchical system of users' needs by Lang and Burnette (1974), and synthesized into the 'habitability framework' by Preiser (1983) and Vischer (1989), among others. Three levels of priority are depicted in Figure 1.2 below. They are:

1. health, safety and security performance;
2. functional, efficiency and work flow performance;
3. psychological, social, cultural and aesthetic performance.

Each category of objective includes sub-goals. At the first level, one sub-goal might be safety; at the second level, a sub-goal can be functionality, effective and efficient work processes, adequate space, and the adjacencies of functionally related areas; and, at the third level, sub-goals include privacy, sensory stimulation, and aesthetic appeal. For a number of sub-goals, performance levels interact. They may also conflict with each other, requiring resolution in order to be effective.

As the three-part Figure 1.2 shows, the three hierarchical levels also parallel the categories of standards and guidelines available to building designers and professionals. Level 1 pertains to building codes and life safety standards projects must comply with. Level 2 refers to the state-of-the-art knowledge about building types and systems, as exemplified by agency-specific design guides or reference works like *Time-Saver Standards: Architectural Design Data* (Watson, Crosbie, and Callender, 1997), or the *Architect's Room Design Data Handbook* (Stitt, 1992). Level 3 pertains to research-based design guidelines, which are less codified, but nevertheless equally important for designers.

This hierarchical system relates the elements of buildings and settings to building users and their needs and expectations. In applying this approach, the physical environment is considered as more than just a building or shell because of the focus on settings and spaces for particular activities engaged in by users. System elements, in effect building performance variables, can be seen as ascending hierarchies from small- to large-scale, or from lower to higher levels of abstraction (see Figure 1.3).

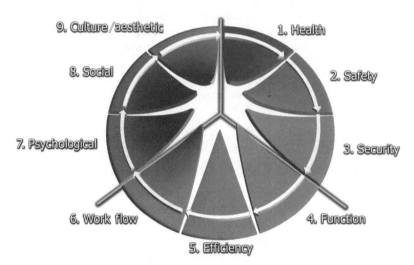

Figure 1.2 Levels of evolving performance criteria.
Source: Jay Yocis, University of Cincinnati.

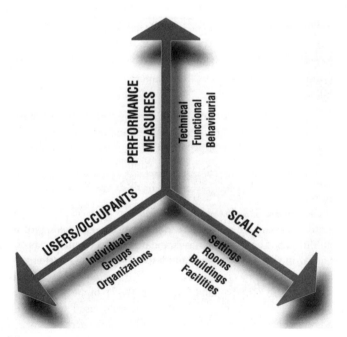

Figure 1.3 Building performance variables.
Source: Wolfgang F.E. Preiser.

For each setting and user group, performance levels for sensory environments and specific quality performance criteria need to be established: e.g. for the acoustic, luminous, olfactory, visual, tactile, thermal, and gravitational environments. While it cannot be sensed by human beings, the effect of electro-magnetic radiation on the health and well-being of people, both from a short-term and a long-term perspective, is also relevant.

In summary, the performance evaluation framework for BPE systematically relates buildings and settings to users and their environmental needs. It represents a conceptual, process-oriented approach that accommodates relational concepts and can be applied to any type of building or environment. This framework can be transformed to permit phased handling of information concerning person-environment relationships; for example, in meeting the need for acoustic privacy at the various phases of programming, developing specifications, building design, and hardware selection.

1.3   Evolving evaluation process models: from POE to BPE

Building performance evaluation is the process of systematically comparing the actual performance of buildings, places and systems to explicitly documented criteria for their expected performance. It is based on the post-occupancy evaluation (POE) process model (see Figure 1.4) developed by Preiser, Rabinowitz, and White (1988). A comprehensive review of POEs at various environmental scales can be found elsewhere (Preiser, 1999; Federal Facilities Council, 2001).

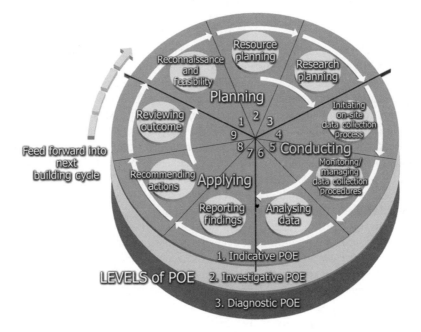

Figure 1.4 Post-occupancy evaluation (POE) process model.
Source: Jay Yocis, University of Cincinnati.

Post-occupancy evaluation (POE), viewed as a sub-process of BPE, can be defined as the act of evaluating buildings in a systematic and rigorous manner after they have been built and occupied for some time. The history of POE started with one-off case study evaluations in the late 1960s, and progressed to system-wide and cross-sectional evaluation efforts in the 1970s and 1980s. Early POEs focused on the residential environment and the design of housing for disenfranchised groups, especially as a result of rapid home construction after the Second World War. Many urban renewal projects in North America, and new town construction in Western Europe, created large quantities of housing without thorough knowledge of the needs, expectations, behaviour or lifestyles of the people they were being built for. The kinds of social and architectural problems that subsequently arose led to an interest in systematic assessment of the physical environments in terms of how people were using them (Vischer, 2001). The approach was later seen as a mechanism for collecting useful information for the building industry on the impact of design and construction decisions over the long term. POEs have since targeted hospitals, prisons, and other public buildings, as well as offices and commercial buildings.

Several types of evaluations are made during the planning, programming, design, construction, and occupancy phases of building delivery. They are often technical evaluations related to questions about materials, engineering or construction of a facility. Examples of these evaluations include structural tests, reviews of load-bearing elements, soil testing, and mechanical systems performance checks, as well as post-construction evaluation (physical inspection) prior to building occupancy. POE research differs from these and technical evaluations in several ways; it addresses the needs, activities, and goals of the people and organizations using a facility, including maintenance, building operations, and design-related questions. Measures used in POEs include indices related to organizational and occupant performance, worker satisfaction and productivity, as well as the measures of building performance referred to above, e.g. acoustic and lighting levels, adequacy of space, spatial relationships, etc.

POE is a useful tool in BPE that has been applied in a variety of situations. In some cases, results are published and widely disseminated; in others, they are uniquely available to the architect, to the client, or to the stakeholder who commissioned the study. The findings from POE studies, while primarily focusing on the experiences of building users, are often relevant to a broad range of building design and management decisions. Many of the building problems identified after occupancy have been found to be systemic: information the engineer did not have about building use; changes that were made after occupancy that the architect did not design for; or facilities staff's failure to understand how to operate building systems.

The BPE framework was developed in order to broaden the basis for POE feedback to include a wider range of stakeholders and decision-makers who influence buildings. This has enabled POEs to be relevant earlier in the design process and applied throughout the building delivery and life cycle. The goal of BPE is to improve the quality of decisions made at every phase of the building life cycle, i.e. from strategic planning to programming, design and construction, all the way to facility management and adaptive reuse. Rather than waiting for the building to be occupied before evaluating building quality, early intervention helps avoid common mistakes caused by insufficient information and inadequate communication among building professionals at different stages.

While POE focused primarily on users' experience of the performance of buildings, the most recent step in the evolution of POE towards building performance evaluation is one that emphasizes a holistic, process-oriented approach toward evaluation. This means that not only facilities, but also the forces that shape them (organizational, political, economic, social, etc.) are taken into account. An example of such process-oriented evaluations is the Activation Process Model and Guide for Hospitals of the Veterans Administration (Preiser, 1997). Process-oriented evaluations are the genesis of BPE and its theoretical framework.

Many stakeholders, in addition to designers and engineers, participate in the creation and use of buildings, including investors, owners, operators, maintenance staff, and, perhaps most importantly, the end users, i.e. the actual persons who occupy and use the building. The term 'evaluation' contains the word 'value', therefore occupant evaluations must state explicitly whose values are invoked when judging building performance. An evalu-ation must also state whose values predominate in the context within which a building's performance is measured. A meaningful evaluation focuses on the values behind the goals and objectives of all stake-holders in the BPE process, in addition to those who carry out the evaluation.

Finally, it should be noted that there are differences between the quantitative and quali-tative aspects of building performance and their respective performance measures, i.e. data that are collected on-site and from building occupants in order to carry out an evaluation. Many aspects of building performance are in fact quantifiable, such as lighting, acoustics, temperature and humidity, durability of materials, amount and distribution of square footage, and so on. Qualitative aspects of building performance pertain to the ambiance of a space, i.e. the appeal to the sensory modes of touching, hearing, smelling, kinesthetic and visual perception, including colour. The qualitative aspects of building performance, such as aes-thetic beauty (i.e. the meaning of buildings and places to people) or visual compatibility with a building's surroundings, can in fact be the subject of consensus among the public. From a planning standpoint, this is evidenced in the process called design review (see Chapter 5), which has resulted in standards for review and guidelines (Scheer and Preiser, 1994). Research consistently shows that the experts and the public disagree on aesthetics and mean-ing, and that while expert decisions do not lead public taste, public opinions have been shown to be stable over time (Nasar, 1999).

1.4   The conceptual basis for BPE

In 1997, the POE process model was developed into an Integrative Framework for Building Performance Evaluation (Preiser and Schramm, 1997), comprising the six major phases of building delivery and life cycle, i.e. planning, programming, design, construction, occupancy and facility management, and adaptive reuse/recycling of facilities. One of the main features in the evolution of this approach was taking into consideration the varying requirements of the process at different points in time, as well as specifying internal review/troubleshooting loops in each of the six phases.

The integrative framework, which is described in detail in Chapter 2, attempts to respect the complex nature of performance evaluation during building delivery as well as during the building's life cycle. This framework is oriented to an architect's perspective, showing the cyclic evolution and refinement of the process while it aims at a moving target: that is, achieving better building performance overall, and better quality as perceived by building occupants (see Figure 2.1 in Chapter 2).

At the centre of the model is the building's actual performance, both measured quanti-tatively and experienced qualitatively. It represents the outcome of the building delivery cycle, and its performance during the life cycle. Each of its six sub-phases has internal reviews and feedback loops. Furthermore, each phase is connected with its respective state-of-the-art knowledge, which is contained in building type-specific databases, as well as in global knowledge and the published literature. For BPE to become viable and truly integrated into the building delivery cycle of mainstream architecture and the construction industry, it is critical to integrate BPE into these disciplines and to demonstrate to practicing professionals the viability of the concept through a range of exemplary case study examples.

1.5   An example of the user feedback cycle in BPE: Building-in-use assessment

In order for the BPE approach to work effectively, data-gathering and analysis activities are necessary at every stage. These can be carried out in a variety of ways including, but not limited to, using traditional social science research techniques.

In view of the fact that the performance criteria at each stage are constituted of both quantitative and qualitative performance evaluation, it is necessary to utilize qualitative and quantitative research. For instance, expected building performance in an area, such as temperature levels inside a building, can be compared with levels of thermal comfort as rated by users. For this comparison to be effective, both the expected and actual perform-ance must use the same or comparable units of measurement. In some fields this can be complicated. For example, expected acoustic performance is usually given in the form of construction materials specifications – distance on centre of wall studs, sound absorption coefficient of ceiling tile. But actual acoustic comfort of users is mostly expressed in the form of a general satisfaction rating, and users' levels of satisfaction or dissatisfaction can only secondarily be compared to measures of expected performance.

One of the challenges of the BPE approach is, therefore, to encourage more precise measures of users' experience of environmental comfort than have conventionally been used. Asking people whether they are satisfied is a rather broad and general outcome measure that tends to include far more than the performance criterion under consideration. Vischer (1989, 1996, 2003) has proposed a technique for approaching users with more direct ques-tions about their comfort levels in relation to various building systems, in order to derive a more specific equivalent to objective environmental measures. The Building-in-use (BIU) assessment system is a validated and reliable standardized survey that can be administered to occupants of any office building in order to collect simple reliable measures of their comfort in regards to seven key environmental conditions. From the responses, scores on the seven comfort conditions can be calculated and compared to a typical or average office building standard derived from a large, user-response database. Thus, deviations from the norm for each of these conditions, in both a positive and a negative direction, provide a quantitative rating of what is essentially a qualitative measure.

The seven conditions addressed by the building-in-use assessment system are: air quality; thermal comfort; spatial comfort; privacy; lighting quality; office noise control and building noise control (Vischer, 1999b). Results of a more recent study indicate that the modern office worker, who must access a variety of equipment and perform a variety of tasks, also evalu-ates comfort in terms of sense of security, building appearance, workstation comfort and overall visual comfort (Vischer et al., 2003). The results of a BIU assessment survey are typically used by facilities managers who must make operating and budgetary decisions; by design professionals who seek to assess an older or about-to-be replaced environment;

in a new building soon after a move; and by building owners and business managers to determine a baseline comfort level for employees in their buildings (Fischer and Vischer, 1998).

One advantage of the BIU assessment system is that it permits comparison between measured and perceived levels of performance. In parts of the building where certain areas of comfort are below the norm, instruments can be applied to measuring conventional performance parameters and determining whether or not standards are being met. User feedback provides a diagnostic data point that permits a wide variety of follow-up actions. However, it is not always possible to determine a direct correspondence between user feedback and the data provided by calibrated measuring instruments (Vischer, 1999a). This is not surprising, in that a number of factors influence the building users, and their experience of one may affect their judgement of another. An important stage of BIU assessment is following up on user feedback by using other measurements to determine performance problems and likely technical causes of low comfort ratings.

There is an ever-increasing variety of diagnostic measuring instruments available for gathering follow-up data on building performance in areas such as indoor air quality and ventilation performance, thermal comfort and humidity, lighting and visual comfort, and noise levels and acoustic comfort. In each of these cases, the researcher must determine how to approach measurement: data-logging over a extended time period, usually in a limited number of places; or spot checks in a compressed time period, but over a larger geographical area. A further issue is how to calibrate the instrument, in terms of its baseline standard or comparator. The human comfort rating is usually the summing up of a wide variety of perceptions and judgements over a given time period; on the other hand, a new situation, or one that causes particular concern to users, may be subjected to short-term judgement that is not indicative of the long-term operation of the building. Finally, no single type of follow-up measurement is mandated: in addition to measuring instruments, evaluators may interview users, question them on psychosocial factors, such as employer-employee relations, and on the requirements of their tasks, which are typically far from uniform.

Other techniques of introducing feedback into the building design and construction process are through checklists, building codes and standards requirements, and design guidelines emanating from other sources. However, this is not to say that in each of these cases some discretionary judgement on the part of the evaluator is not required to ensure that the process moves forward. For data-gathering techniques in BPE to be valid and standardized, the results need to become replicable. The ultimate goal of the International Building Performance Evaluation (IBPE) Consortium (Preiser, 1995) is to create a standardized 'universal tool kit' of data-gathering instruments, which can be applied to any building type anywhere in the world (Preiser and Schramm, 2002). A preliminary 'tool kit' is therefore provided in the Appendix to this book.

1.6   Economic and sustainability issues

Since a major concern of practitioners in the construction and real estate industry is the cost of time, innovative techniques that could be applied to programming, design and construction are often considered too costly for a typical building project's time-frame and budget. As a result, proposals to incorporate additional information in the form of user feedback or sustainability recommendations, especially after a project has started, are typically rejected by project managers, even if they may improve the outcome. However, shortcut methods, such as BIU assessment, Serviceability assessment (see Chapter 10) and the Balanced scorecard approach (see other examples in Chapter 9), have been devised to allow researchers/evaluators to obtain valid and useful information in a much shorter time-frame than was previously possible. Understanding the BPE sequence enables them to apply information quickly and at appropriate points in the process. This enables evaluation to become cost-effective and to respect time constraints.

Each information loop in the BPE process offers an opportunity for incorporating environmental and sustainability concerns into building design and construction. Normally viewed as an 'add-on' in conventional building projects, environmentally responsible technology and materials can be a financially viable alternative if they are introduced as options from the beginning and consistently included throughout the process. By systematically collecting feedback and using the life cycle perspective, the right information is available at the right time for each key decision, thus increasing building quality without affecting project time and costs adversely.

1.7   Conclusions

Criteria for designing and building new environments should be based on the evaluation of existing ones, and modified when appropriate in the context of the design process. The crucial point in every evaluation is to identify the significance of results and how best to apply them constructively. Product evaluation and quality control, in terms of product performance and customer satisfaction, is an accepted procedure in most industries. Product improvement is ongoing and forms an integral part of the price calculation. It would seem both natural and economical in the long run to thoroughly investigate the possibilities of user-oriented 'product' evaluations in the complex industry of design and construction.

While the constructed environment cannot be called a 'product' in the strictest sense, because it is dynamic and changing over time, feedback and evaluation costs can be included in the life cycle costing of buildings as an 'extra' that might be written off against taxes as an investment in the future. Building the costs of building evaluation into construction projects would enable large-scale environmental research to be funded, and thereby launch an ongoing national evaluation research programme. Demographic change and an aging population place new requirements on such sectors as housing in urban and non-urban areas, including the large retirement communities of the 'Sunbelt' states in the USA. Applying the BPE framework to large-scale residential construction could not only improve the cost and quality of such housing, but it would also ensure that the environments occupied by these users meet criteria of environmental quality, cost-effective construction practices, and other social needs.

In the remaining chapters of this book, architects, planners, programmers, and researchers elaborate on the different phases of BPE, and provide case study examples of how some of the different stages have been implemented. The first phase is strategic planning, and the review loop case study example demonstrates effectiveness review; the second phase is programming, also known as briefing, with a programme review loop providing examples of how to complete this critical stage properly. The design phase has design review as its loop, and an example is presented of major renovations at a manufacturing company. This is followed by the construction phase and its commissioning review loop, an important process in quality assurance in buildings. Once the building is occupied, concerns with quality become even more important. Later chapters address the difficulties of routinely implementing POE in managed buildings and explain how the BPE approach is a tool at several different levels for facilities managers, whose responsibilities for management and maintaining quality in buildings range from technical maintenance to user satisfaction to ecological sustainability.

In Part Two, chapters discuss various approaches to BPE in other cultural contexts, with examples of methodologies and approaches from Brazil, Canada, Germany, Israel, Japan, and the Netherlands. In addition, BPE is linked to universal design and facility performance evaluation, as well as to responsible humanistic decision-making and sustainability. Finally, two chapters outline ways of ensuring that innovative technology, as well as innovative user participation processes, are integrated into the BPE approach.

Learning and appreciating how to apply the conceptual framework for building performance evaluation to future building design, construction and operation is an exciting new frontier in an industry that is slow to embrace change. The energy and enthusiasm of practitioners and researchers in this field are creating a wealth of data and knowledge that it is no longer possible to ignore. It is comforting to contemplate that as traditional building practices change, so our buildings will become more comfortable and more humane, as well as more environmentally responsible. This book is a critically important step in this direction.

References

Conan, M. (1990). *Critical Approaches to Environmental Design Evaluation, 2 vols.* Conference proceedings, Centre Scientifique et Technique du Bâtiment: Paris, July.

Federal Facilities Council (2001). *Learning from our Buildings: A State-of-the-Practice Summary of Post Occupancy Evaluation.* National Acadamies Press, Washington, DC.

Fischer, G.N. and Vischer, J.C. (1998). *L'évaluation des environnements de travail, la méthode diagnostique.* Montréal: Presses de l'Université de Montréal and Brussels: DuBoek.

Foerster, H., von (1985). *Epistemology and Cybernetics: Review and Preview* (Milan, Italy: Lecture at Casa della Cultura, 18 February).

Infoplease Dictionary (2003). Web based definition of cybernetics.

Lang, J. and Burnette, C. (1974). A Model of the Designing Process. In *Designing for Human Behavior* (W. Moleski and D. Vachon, eds), pp. 43–51. Dowden Hutchinson and Ross.

Maslow, H. (1948). A Theory of Motivation. *Psychological Review*, **50**, 370–398.

Nasar, J.L. (ed.) (1999). *Design by Competition: Making Design Competition Work.* Cambridge University Press.

National Council of Architectural Registration Boards (NCARB) (2003). *Improving Building Performance.*

Preiser, W.F.E. (1983). The Habitability Framework: A Conceptual Approach Toward Linking Human Behavior and Physical Environment. *Design Studies*, Vol. 4, **2**, April.

Preiser, W.F.E. (1989). *Building Evaluation.* Plenum.

Preiser, W.F.E. (1991). Design Intervention and the Challenge of Change. In *Design Intervention: Toward a More Humane Architecture* (W.F.E. Preiser, J.C. Vischer, and E.T. White, eds). Van Nostrand Reinhold.

Preiser, W.F.E. (1997). Hospital Activation: Towards a Process Model. *Facilities*, **12/13**, 306–315, December.

Preiser, W.F.E. (1999). Post-Occupancy Evaluation: Conceptual Basis, Benefits and Uses. In *Classical Readings in Architecture* (J.M. Stein and K.F. Spreckelmeyer, eds). McGraw-Hill.

Preiser, W.F.E. (2001). The Evolution of Post-Occupancy Evaluation: Toward Building Performance and Universal Design Evaluation. In *Learning from our Buildings: A State-of-the-Practice*

Summary of Post-Occupancy Evaluation (Federal Facilities Council). National Academies Press, Washington, DC.

Preiser, W.F.E. and Schramm, U. (1997). Building Performance Evaluation. In *Time-Saver Standards: Architectural Design Data* (D. Watson et al., eds). McGraw-Hill.

Preiser, W.F.E. and Schramm, U. (2002). Intelligent Office Building Performance Evaluation. *Facilities*, Vol. 20, **7/8**, 279–287.

Preiser, W.F.E., Rabinowitz, H.Z. and White, E.T. (1988). *Post-Occupancy Evaluation.* Van Nostrand Reinhold.

Scheer, B.C. and Preiser, W.F.E. (eds) (1994). *Design Review: Challenging Urban Aesthetic Control.* Chapman & Hall.

Stitt, F.A. (1992). *Architect's Room Design Data Handbook.* Van Nostrand Reinhold.

Vischer, J.C. (1989). *Environmental Quality in Offices.* Van Nostrand Reinhold.

Vischer, J.C. (1996). *Workspace Strategies: Environment As A Tool For Work.* Chapman & Hall.

Vischer, J.C. (1999a). Using Feedback From Occupants to Monitor Indoor Air Quality. *Proceedings IAQ93, ASHRAE*: Denver, June 1993.

Vischer, J.C. (1999b). A Useful Tool for Office Space Evaluation. *Canadian Facilities Management*, September, p. 34.

Vischer, J.C. (2001). Post-Occupancy Evaluation: A Multi-facetted Tool For Building Improvement. In *Learning from our Buildings: A State-of-the-Practice Summary of Post-Occupancy Evaluation* (Federal Facilities Council). National Academy Press.

Vischer, J.C. et al. (2003). *Mission impossible ou mission accomplie? Évaluation du mobilier universel chez Desjardins Sécurité Financière, 2 vols.* Technical report: Groupe de recherche sur les espaces de travail, Université de Montréal.

Vitruvius, (1960). *The Ten Books on Architecture* (translated by M.H. Morgan). Dover Publications.

Watson, D., Crosbie, M.J. and Callender, J.H. (eds) (1997). *Time-Saver Standards: Architectural Design Data.* McGraw-Hill (7th Edition).

2

A conceptual framework for building performance evaluation

Wolfgang F.E. Preiser and Ulrich Schramm

Editorial comment

Building performance evaluation (BPE), as conceptualized by the authors, grew out of post-occupancy evaluation (POE), an established research method to evaluate buildings at different levels of effort and sophistication after they are occupied. According to the editors' search of the literature, the first publication with the term 'post-occupancy evaluation' in its title goes back to the 1970s (McLaughlin, 1975). From the specific focus on this phase of building occupancy, as explained in Chapter 7, the POE process model was expanded into an integrative framework for building performance evaluation (Preiser and Schramm, 1997). An important feature of this framework was the time dimension, which took into consideration the complex nature of performance evaluation in the building delivery cycle, as well as the entire life cycle of buildings. The six phases of the 1997 'integrative framework for BPE' are: (1) strategic planning, (2) programming, (3) design, (4) construction, (5) occupancy, (6) adaptive reuse/recycling.

This chapter describes the six phases and the internal review loops of each one. As the conceptual framework for BPE has evolved, the names of some of these phases and their review loops have been modified. Going forward, the application of the 'integrative framework for BPE' may increasingly be led by Facilities Management (FM). This chapter will shed light on the goals, potential, and benefits of applying BPE over the lifetime of a building. The ultimate purpose is to inspire respect for all participants involved in building delivery and use to create a common understanding of the interdependencies between all phases of the building life cycle, and to achieve better performing buildings in the future.

2.1   Introduction

The BPE framework draws on a model of continuous quality improvement to encompass the design and technical performance of buildings, and to contribute to knowledge-building in the design and construction industry. This comprehensive approach to building performance evaluation is applicable to all facility types.

For a given building type, location and cultural context, the expected performance of the building needs to be defined and communicated to those who programme, design, and, ultimately, operate the facility. It is important to remember that the physical and technical performance of buildings is directly linked to the building qualities perceived by occupants. That is to say, occupants' perceptions are as significant as those building attributes that are defined by independent measures when a building is evaluated. A design has to be evaluated according to how it is used and not on how it appears to the designer. For example, staircases that are poorly lit, have poorly differentiated colours and materials for risers and treads, or even with a distracting view, are frequently cited for causing accidental slips and falls, regardless of whether or not users have vision or mobility problems. In such cases as this, evaluation requires an understanding of both the physical aspects of the design, and how it is perceived by users. The staircase example demonstrates how the physical, technical and behavioural performance of buildings – involving quantitative and qualitative measures – are inextricably linked to evaluation.

In the sections that follow, the six phases of building performance evaluation (BPE) are presented as categories for specifying expected quantitative and qualitative performance scales for different types of built environments. These are based on types and numbers of expected users, space-use patterns, health, safety and security criteria, functional criteria, social, psychological and cultural criteria, ambient environmental conditions, spatial relationships, equipment criteria, code criteria, special requirements, and last, but not least, estimated space needs (Preiser, Rabinowitz, and White, 1988). BPE constitutes an important step in validating performance standards that may already exist, or that have to be developed for a given building type.

2.2   Description of the conceptual framework for BPE

While in the past building delivery was seen as a linear, end product-oriented process, the integrative framework is a dynamic, evolving and non-mechanical model (Petzinger, 1999), which can be depicted as an ever-expanding helix of knowledge on building performance. As stated above, it attempts to respect the complex nature of performance evaluation in the building delivery process, as well as throughout the entire life cycle of buildings. The BPE framework defines the building delivery and life cycle from the perspective of all parties who are involved with a building. Over time, by focusing on 'data capture' from repeat evaluations of buildings, it is expected that knowledge about building performance will be accumulated in building type-specific databases and information clearinghouses (see Figure 2.1).

Quantitative and qualitative building performance criteria that represent the expected outcome or product of the building delivery process, as well as building performance during its life cycle, are at the centre of the model. It shows the six sub-phases, each of which has internal review and feedback loops and is connected with state-of-the-art knowledge contained in building type-specific databases, published guidelines, as well as proprietary

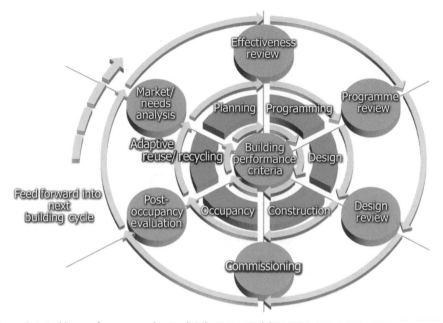

Figure 2.1 Building performance evaluation (BPE) process model.
Source: Jay Yocis, University of Cincinnati.

expertise residing in specialized firms, for instance. The phases and feedback loops of the framework are described below:

Phase 1 – strategic planning: the starting point of the building delivery cycle is the strategic plan, which establishes medium- and long-term needs of an organization through market/ needs analysis and, in turn, is based on project mission and goals, as well as facility audits. If, for example, the statement, 'being close to the customer' is part of a global organization's mission, the market for its products has to be analysed in order to identify possible locations for regional headquarters or subsidiaries (Petronis, 1993). Facility audits identify needs, such as space, and compare them with existing resources in order to establish demand.

Loop 1 – effectiveness review: outcomes of strategic planning are reviewed in terms of their effectiveness, relating to the specific 'big issue' categories of a given organization that match its mission and goals, such as corporate symbolism and image, visibility, innovative technology, flexibility and adaptive reuse, initial capital cost, operating and maintenance cost, and costs of replacement and recycling. Chapter 3 elaborates on Phase 1 and its feedback loop.

Phase 2 – programming: once strategic planning, cost estimating and budgeting has occurred, a building project is a reality and programming ('briefing' in the UK) can begin. Over the past 30 years, programming has become a required step in building delivery. It is outlined in the standard American Institute of Architects (AIA) manual and described by the International Organization for Standardization as an international standard (ISO 9699). A programme is necessary when the client and/or the future building user are attempting, in consultation with building specialists, to document the needs, aims, resources and the context of the project, and solutions to other problems arising at this phase.

Loop 2 – programme review: at the end of this phase, programme review involves the client, the programmer, and representatives of occupant groups. This allows the project participants to reflect on the programme document containing performance criteria and other outcomes of strategic planning. The review process allows the programme to be evaluated step-by-step and to be modified in response to requirements or new priorities that might have emerged as part of the planning and programming process. The development of the programme should be a dynamic process, whereby team creativity and systematic problem analysis combine to resolve conflicts in this phase. Chapter 4 gives a detailed description of Phase 2 and its feedback loop.

Phase 3 – design: this phase includes schematic design, design development and construction documents. In the design process, schematic design is the initial phase of building design, during which a range of alternative solutions are developed, translating the programming parameters into one or more broad-brush building solutions. Design development is the second stage of building design, wherein one of the alternatives is chosen and elaborated on in design development, addressing in detail more of the issues raised by the programme. Finally, construction documents are produced for the design solution chosen. In this step, all relevant information is synthesized into the practical instructions and requirements needed to build the facility. This includes technical information about building materials, equipment, furnishings, and systems.

Loop 3 – design review: the design phase has evaluative loops in the form of design review, or 'troubleshooting', involving the architect, the programmer, and client or user representatives. The development of knowledge-based and computer-aided design (CAD) techniques makes it possible to evaluate solutions during the earliest phases of design. This allows designers to consider the effects of design decisions from various perspectives, while it is still not too late to make modifications.

The formal or stylistic aspects of the building solution are evaluated differently by clients, users, architectural critics, and other architects. However, a well-written and illustrated programme can serve as the basis for a more objective evaluation of the design by using, for example, bubble diagrams, vignettes or situational sketches, as well as images of comparable and exemplary solutions. The goals of the organization, as well as its specific programme requirements, provide evaluation criteria against which programmer, client, and users can judge the building design as it develops. In areas where the design fails to meet programme requirements, the client has to decide if the relevant programme parameters, such as budget or functionality, need to be modified. Such changes need to be documented and added to the programme. This also holds true if the client's programme requirements change as a result of the ongoing and dynamic process of building delivery. Programme tracking is therefore essential in order for subsequent, programme-based design review to be realistic, useful and effective. Chapter 5 elaborates on Phase 3, design, and its review loop.

Phase 4 – construction: once design review has occurred, building construction can begin. The programme, working drawings and written construction documents are all part of the building contract, and they describe the expected performance of the future building in detail. In this phase, construction managers and architects share in construction administration and quality control to assure contractual compliance. In addition, national standards and codes as well as local regulations need to be met, including quality standards or safety regulations. Failure to complete the previous phases can result in unforeseen 'change orders' during construction, as some new requirement is identified or budgetary constraints imposed. Responding to change orders can substantially alter the cost of building construction, while value engineering can help reduce costs significantly.

Loop 4 – commissioning: at the end of the construction phase, inspections take place, which result in 'punch lists'; that is, items that need to be completed prior to acceptance and occupancy of the building by the client. As a formal and systematic review process, this loop is intended to insure that owners' expectations, as well as obligatory standards and norms, are met in the constructed building. This feedback loop is a 'reality check': it ensures that the builder fulfils his contract and that specific building performance criteria are made explicit, as well as compliance with relevant standards and norms. Chapter 6, on building commissioning, explains the construction phase and feedback loop.

Phase 5 – occupancy: in temporal terms, this phase is the longest of all those described. In fact, most analytic approaches to building delivery end at move-in. However, the BPE approach, with its reliance on feedback and evaluation, maintains a long-term perspective by including the period of occupancy in order to improve the quality of decisions made during the earlier phases. While the earlier sub-phases normally last a couple of months, occupancy may last 30–50 years, depending on building type. To occupy a building is the original goal of a client when they decide on a building project. Although strategic planning, facility programming and design are important phases in the quest to realize a building, it is only at move-in that the client obtains the architectural solution to the initial problem. During this phase, there is fine-tuning of the environment by adjusting the building and its systems to achieve optimal functioning for occupants.

Loop 5 – post-occupancy evaluation: during this phase, BPE is activated in the form of POEs that provide feedback from users on what works in the facility and what needs improvement. POEs also test some of the hypotheses behind key decisions made in the programming and design phases. Alternatively, POE results can be used to identify issues and problems in the performance of occupied buildings, and identify ways to solve these. Moreover, POEs are ideally carried out at regular intervals, that is, in two- to five-year cycles, especially in organizations with repetitive building programmes, such as school districts and federal government agencies. The occupancy phase and its feedback loop are covered in Chapter 7.

Phase 6 – adaptive reuse/recycling: in many cases, recycling buildings for similar or quite different uses towards the end of their useful life has become quite common. Lofts have been converted to artist studios and apartments; railway stations have been transformed into museums of various kinds; office buildings have been turned into hotels; and factory space has been remodelled into offices or educational facilities. Such major use changes are as dramatic as constructing a new building. Even if building use does not change, building interiors are changing constantly throughout the lifetime of a building. The question of how well a building adapts and can be recycled, not only in the sense of sustainable building practices, but also in the sense of adaptation to new uses, is highlighted in Chapter 9. The end of this phase constitutes the end of the useful life of a building, when the building is decommissioned and mothballed or demolished. In cases where construction and demolition waste reduction practices are in place, building materials with potential for reuse will be sorted and recycled into new products and hazardous materials removed.

Loop 6 – market/needs analysis: this loop involves evaluating the market for the building type in question in terms of the client organization's needs. It can mean assessing the rehabilitation potential of an abandoned or stripped-down building shell, or the potential of a prospective site in terms of future needs. Thus, in the BPE framework, the end point of this evolutionary cycle is also the beginning point of the next building delivery cycle. This phase and feedback loop is discussed in Chapter 8 on facilities management.

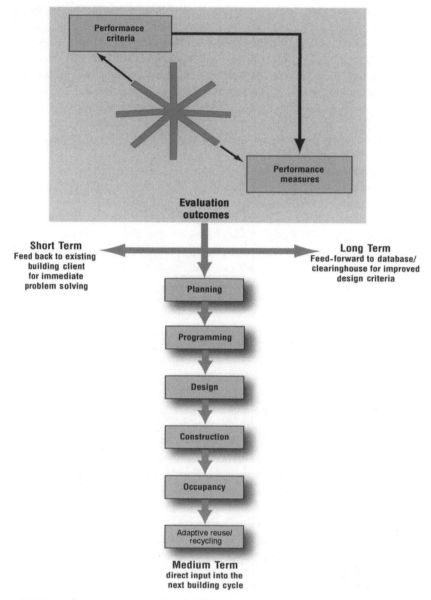

Figure 2.2 The performance concept and the building process.
Source: Architectural Research Consultants, Albuquerque, NM.

2.3   The performance concept and the building process

The 'Performance Concept', outlined in Chapter 1, evolved in the 1960s, and is the basis for the evaluation framework presented here (Figure 2.2). Throughout history, building performance has been evaluated in an informal manner. The lessons learned may or may not have been applied in the next building. Because of relatively slow change in the evolution of building types in the past, knowledge about their performance could be passed on from generation to generation of building specialists. These were often craftspeople with multiple skills – artists/designers/draftsmen/builders – who had more control over the building delivery process than the members of multi-disciplinary teams that are typically involved in building today. The emergence of new professions means increasing specialization; for example, project management focuses primarily on the building delivery process, and facility management focuses on operating the building over its lifetime. In order to achieve a successful building project, all those involved need strong communication skills, as explained in Chapter 18, in order to integrate the different specialized areas of knowledge.

Today there is increasing specialization not only in the construction industry, but also in the demands clients place on facilities. The situation is made more difficult by the fact that no one person or group controls the building delivery process. Rather, major building decisions tend to be made by committees, while an increasing number of technical and regulatory requirements are imposed on building projects, such as access for the disabled, energy conservation, hazardous waste disposal, fire safety, and occupational health and safety requirements. As a result, building performance criteria need to be well articulated and documented, usually in the form of the programme or brief, and applied throughout the process.

2.3.1   Outcomes

Figure 2.2 illustrates the performance concept in the building process, as well as outcomes from a short-, medium- and long-term perspective. As indicated in Chapter 1, the concept is shown as a basic feedback system comparing explicitly stated performance criteria with the actual, measured performance of a building. This comparison – the core of the evaluation process – implies that expected performance can be clearly expressed as performance criteria. Programming is the process of systematically collecting, documenting, and communicating the criteria for the expected performance of a facility, and POE is the review loop at occupancy. Building performance is systematically compared with the expected performance criteria documented in the facility programme.

The figure shows that the outcome of a building performance evaluation varies according to short-, medium- and long-term time-frames. Short-term outcomes include user feedback on problems in building performance within a specific sub-phase of the building life cycle, and identification of appropriate solutions. Medium-term outcomes include applying the positive and negative lessons learned to inform subsequent phases within a building's life cycle, as well as the next building delivery cycle. Long-term outcomes are aimed at the creation of databases, clearinghouses and the generation of planning and design criteria for specific building types. Database development, such as those listed in Section 2.3.3, below, assumes a critical role in linking various review loops like post-occupancy evaluation, with planning and programming.

2.3.2   Performance levels and building performance variables in BPE

The performance levels and building performance variables described earlier (see Chapter 1) speak to the interrelationships that exist between the built environment, service providers,

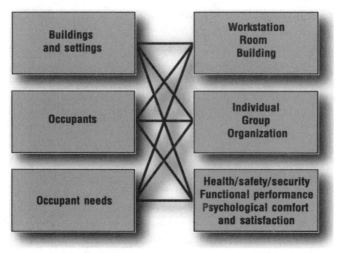

Figure 2.3 Interrelationships between buildings, occupants and occupant needs. *Source:* Wolfgang F.E. Preiser.

and users, as well as to client/user goals and needs. Each of the categories is presented in Figure 2.3.

- Built environment: workstations, rooms, buildings, and entire complexes of buildings or facilities.
- Providers and users: individuals, groups, and entire organizations.
- Performance levels and criteria: based on client goals and user needs, this hierarchy of performance levels includes technical (health, safety, security), functional (functionality, efficiency, work flow), behavioural (social, psychological, cultural), and aesthetic performance criteria.

Not shown in Figure 2.3 are contextual variables. The well-established categories listed above are embedded in a fourth, overarching category at the global or 'meta' level of context-related aspects. They are process-driven, and they deal with overall vision, as well as with historical, political, socio-economic, cultural, and other significant aspects. An excellent example of such an approach is the successfully redeveloped area across from Tokyo Station in Japan. The Marunouchi building by Mitsubishi is a mixed-use tower with much of the space given over to public use, in addition to the usual office, commercial, restaurant and meeting spaces (see Figure 2.4). It was fully leased shortly after opening in 2002, although there was a glut of vacant office space in Tokyo at the time. Since then, it has become the driving force for revitalization of the surrounding area.

2.3.3 Development of performance criteria

For BPE to be objective, actual performance of buildings is measured against established performance criteria. There are several sources for such criteria:

- Published literature: published literature can provide explicit or implicit evaluation criteria. Explicit criteria are contained in reference works and publications. Implicit criteria can be derived from research journals or conference proceedings, which contain the

(a)

Figure 2.4a & b Successful mixed-use concept: the Marunouchi building in Tokyo, Japan.
Source: Mitsubishi.

findings and recommendations of research linking the built environment and people. Implicit criteria require interpretation and validation by the evaluator, as data need to be assessed for appropriateness to a given context.

● Analogues and precedents: in cases where a new building type is being programmed and designed, performance criteria need to be compiled for spaces and buildings for which no precedents exist. In these cases, the most expedient method for obtaining performance criteria is to use so-called analogues; that is, to 'borrow' criteria from similar, but not identical, space types, and to use educated guesses to adapt them to the situation at hand. For example, in designing new centres for non-intrusive cancer treatments in the 1980s, the question had to be asked: Should it be a free-standing facility, or should it be integrated into a major hospital? Should the scale be non-institutional and the ambiance home-like, as opposed to a cold, institutional atmosphere? These questions led to designs intended to create a 'de-institutionalized' and humane environment of modest scale, with an empathetic ambiance, which would help reduce the stress for cancer

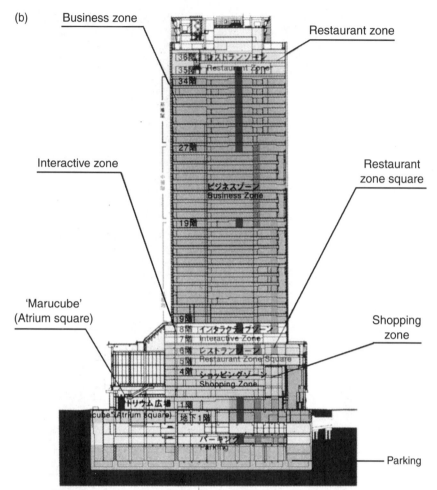

Figure 2.4a & b (Continued).

patients and families. Similarly, when residential and day-care facilities for Alzheimer's patients were first designed, little knowledge existed as to which performance criteria would be appropriate for patients at different stages of the disease. Through design research, a number of hypotheses about patients' needs and abilities – such as wayfinding, socializing, and daily activities – were tested and have subsequently found their way into the design guidelines (Cohen and Weisman, 1991).

● BPEs/POEs: the third source of criteria development is performance evaluation and feedback through BPEs/POEs, described in detail in Chapter 7. Key design concepts, building typology, and the actual operations of facilities are evaluated. Feedback can be generated by specialist consultants, for example, or by facilities managers who monitor certain building performance aspects on a daily basis. Evaluation includes the performance of materials and finishes, their maintainability and durability, cost of replacement, and frequency of repair. This also applies to the performance of hardware, such as window systems and locking systems, as well as heating and air-conditioning systems. Feedback adds contextual information to state-of-the-art knowledge for a given building type, such as office buildings (Preiser and Schramm, 2002).

- Resident experts: a fourth source of performance criteria are knowledgeable people or 'resident experts'; that is, people familiar with the operation of the facility in question. People who are experienced operators of a facility are likely to represent informed judgement and experience. To help elicit such expertise, focus group sessions can be used to discuss advantages or disadvantages of certain performance aspects of the facility type in question. Furthermore, representative user group and consensus discussion can be used to generate performance criteria that are appropriate for the tasks at hand.

2.3.4   Evaluation methods

Major contributions to the field of building evaluation are identified by Preiser, Rabinowitz, and White (1988), Preiser (1989), and Baird et al. (1996). These books offer state-of-the-art methods for monitoring and understanding building performance that can be used worldwide. The methods explore the 'demand' side, that is, occupant requirements, as well as the 'supply' side, that is, the building's capabilities to meet these requirements. Although there are some similarities among methods in terms of providing systematic ways of measuring the quality of buildings, they differ in their target audiences, their scope, and systems of measurement.

Several observations can be made about the current status of evaluation methods and their representation in the chapters that follow:

- Evaluation efforts started with fairly modest and typical singular case studies (Counell and Ostrander, 1976). Over the years, the level of sophistication has increased and greater validity of data has been witnessed (Federal Facilities Council, 2001).
- Standardization is beginning to be employed in evaluation methodology in the quest for making evaluations replicable and more generalized (see Chapters 1 and 10).
- In the context of globalization, there is a special need for an increased understanding of inter-cultural differences in the use of evaluation methods and techniques (Schramm, 1998).
- User involvement is one of the major characteristics of many evaluation methods. For an organization, involving users is a credible expression of an open culture and humane values, while for users it is a big step towards empowerment (see Chapter 5).
- A multi-method approach is critical to enhance the credibility of findings. Good programming, for example, includes various methods such as review of case studies, focus groups, questionnaires and diaries (see Chapter 4).
- The evaluator is the driving force of the evaluation system. He or she needs to act as a facilitator to ensure that all interest groups have a chance to express their views and to monitor the process in a constructive way (see Chapters 5 and 18).
- In addition to traditional tools, the development of evaluation methods and techniques is an ongoing process, especially in light of innovations in information technology, which permit specific hardware and software development (see Chapters 7 and 14).
- All evaluation methods have their strengths and limitations. The pros and cons of methods need to be weighed carefully depending on the phase within the building life cycle and the political and economic context.

2.4   Conclusions

In the context of increasing globalization of business and institutions, the role BPE plays is growing in significance. On one hand, this trend leads to standardization of certain building types and their floor plans, e.g. hotel 'brands' tend to conform to worldwide standards of room sizes, floor plans, and amenities. On the other hand, social and cultural differences and varied building traditions require differentiation in design which is adapted to context, e.g. for buildings such as community centres, museums, or other local, cultural institutions.

Building performance evaluation identifies both successes and failures in building performance by placing an emphasis on human factors and on the interaction between designed physical settings and building systems. Using the BPE approach as standard practice will help professionals to establish a performance-based approach to building design and an evaluative framework that improves outcomes throughout the design and building delivery cycle, extending through the life cycle of the building. Benefits include better quality of the built environment; greater occupant comfort and a more satisfactory experience in visiting, using, or working in a facility; improved staff morale; improved productivity; and significant cost savings. Most important of all, building performance evaluation contributes to the state-of-the-art knowledge of environmental design research and thus makes significant contributions towards improving the profession of architecture and increasing quality in the building industry.

References

Baird, G., Gray, J., Isaacs, N., Kernohan, D. and McIndoe, G. (eds) (1996). *Building Evaluation Techniques*. McGraw-Hill.

Cohen, U. and Weisman, G.D. (1991). *Holding on to Home: Designing Environments for People with Dementia.* The Johns Hopkins University Press.

Connell, B.R. and Ostrander, E.R. (1976). *Methodological Considerations in Post Occupancy Evaluation: An Appraisal of the State of the Art*. Washington, DC: The American Institute of Architects Research Corporation.

Federal Facilities Council (2001). *Learning from Our Buildings: A State-of-the-Art Practice Summary of Post-Occupancy Evaluation.* National Academy Press.

McLaughlin, H. (1975). Post-occupancy evaluation of hospitals. *AIA Journal*, January, pp. 30–34.

Petronis, J.P. (1993). Strategic Asset Management: An Expanded Role for Facility Programmers. In *Professional Practice in Facility Programming* (W.F.E. Preiser, ed.). McGraw-Hill.

Petzinger, T. (1999). A New Model for the Nature of Business: It's Alive! *The Wall Street Journal*, February 26.

Preiser, W.F.E. (ed.) (1989). *Building Evaluation*. Plenum.

Preiser, W.F.E. and Schramm, U. (1997). Building performance evaluation. In *Time-Saver Standards for Architectural Data* (D. Watson et al., eds), pp. 233–238. McGraw-Hill.

Preiser, W.F.E. and Schramm, U. (2002). Intelligent Office Building Performance Evaluation. *Facilities*, Vol. 20, **7/8**, pp. 279–287.

Preiser, W.F.E., Rabinowitz, H.Z. and White, E.T. (1988). *Post-Occupancy Evaluation*. Van Nostrand Reinhold. (Out of print: contact <wolfgang.preiser@uc.edu> regarding availability of copies.)

Schramm, U. (1998). Learning from Building-User Feedback: The Post-Occupancy Evaluation Process Model in the Cross-Cultural Context. In *Architecture and Teaching: Epistemological Foundations* (H. Dunin-Woyseth and K. Noschis, eds), pp. 127–133. Chabloz.

PART TWO

Performance Assessments in the Six-Phase Building Delivery and Life Cycle

3

Phase 1: Strategic planning – effectiveness review

Ulrich Schramm

3.1   What is strategic planning?

The improvement of living conditions, socially and economically, as well as the protection of natural resources, are goals of sustainable development, and can be applied specifically to the building and construction industries with their complex impact on the natural environment. In order to strengthen sustainability within the building and construction industries, so-called 'win-win' situations need to be generated, with ecological, economical and social benefits for stakeholders. To this end, a holistic view of a building's life cycle must begin as early as possible, that is, when building-related decisions become part of an organization's general strategy.

A strategy is a careful plan or method for achieving a specific end (*Merriam-Webster's Collegiate Dictionary*, 11th Edition, 2003). Thus, strategic planning is a process that culminates in a strategic plan based on the medium- and long-term needs of an organization as determined by a market/needs analysis. This, in turn, is based on the organization's mission and goals, as well as facility audits (Preiser and Schramm, 1997).

At this early stage of strategic planning, a building project does not have a clear shape or form. Based on market analysis and statement of needs, there may be several options within the corporation's strategy. The range of these options may be diverse in terms of scale, resources, actions, effects, etc. At one end of the spectrum, an option may respond to identified needs with a 'bundle' of various resources, such as organizational, process/manpower-related, infrastructural, etc. At the other extreme, the option may be 'to do nothing'. Between these two extremes, options may include: addressing environmental aspects at different scales related to the technical infrastructure in general; selecting facilities; leasing in existing office buildings; or even planning new construction. As part of an overall strategic plan, an institution may use current assets, such as cash, to create or maintain non-current assets, i.e. infrastructure, buildings, land, and major equipment (Petronis, 1993).

As a result of the strategic plan, several building projects could be initiated, or only one, or none at all.

Consequently, in the first phase of building performance evaluation, known as 'strategic planning', the focus is on options, including first ideas, general concepts, or stimulating scenarios, rather than on well-defined building projects with specific facility programmes. In contrast, the focus of programming (Phase 2) is typically on defining design requirements for specific facility projects (Petronis, 1993). Programmers are cognizant of the economic consequences of their recommendations, and the focus of their efforts is generally on the scale of a single building complex. However, it is also important to recognize the larger institutional decision-making framework in which first strategic planning and later, facility programming, occur.

In the UK, the term 'programming' or 'briefing' refers to a building project that will help to achieve the overall goals already established (see Chapter 4). The programme or brief describes the parameters or performance requirements to be fulfilled by the design solution, i.e. the resulting building, whereas the term 'strategic planning' is the activity of the corporation prior to embarking on a building project. Strategic planning is a planning activity to enable the corporation to decide where to go and what to achieve within a given framework of time and money, e.g. within a five-year-plan. The charge is to set goals and objectives according to the business needs identified through market analysis. Sometimes 'strategic programming' or 'strategic briefing' is used in the literature as a synonym for strategic planning (Blythe and Worthington, 2001). However, in the context of building performance evaluation, 'strategic programming' is a necessary precursor to programming. It may have one or more building projects as its focus, and it may be performed by facility programmers with either an emphasis on basic quantitative parameters (Kumlin, 1995) or on architectural programming (Hershberger, 1999).

Just as programming for a new building should be as extensive and inclusive as possible, strategic planning must address many levels of decision-making. As a result, there are different programming methods with many variations, and different levels of sophistication within each method, which can also be applied to strategic planning. Fourth-level programming, for example, includes political considerations, i.e. where planning problems are commingled with political issues and power struggles (Pena et al., 1987).

3.2   Why strategic planning?

It is evident that strategic planning never starts from zero: there is always a status-quo situation with which members of an organization may or may not be satisfied. This existing situation can then be analysed in order to set a direction, such as defining a specific building project in the context of long-term strategic goals. Once a project has been defined and agreed upon, programming may begin. Most experts writing in this area agree that some strategic planning by the client organization is required before programming or briefing can start. Kumlin, for example, concludes that sometimes the idea for the creation of a new facility is the result of new insight at the strategic level that precedes the programming process (Kumlin, 1995). Petronis, as stated above, acknowledges that facility programming itself is an important activity and stresses the importance of recognizing that programming exists within a larger planning perspective (Petronis, 1993). Finally, Hershberger argues that three different types of planning studies may be required before architectural programming commences: financial feasibility; site suitability; and master planning (Hershberger, 1999). As part of strategic planning, he stresses the need for a financial feasibility study. This study may be conducted even before a site has been selected. It involves predicting if the market conditions, available financing, site situation, and building costs will converge in such a way as to lead to a successful project – one that will provide a favourable return on investment. For him, it makes little sense to do architectural programming for a facility until it is certain that what is being proposed is economically viable.

Financial constraints, as well as time constraints, are only some of the issues that need to be clarified during strategic planning before facility programming can start. Other issues are 'laws, standards and codes' or 'client's future enterprise' (see ISO 9699, Annex B.2 and B.6, 1994; DIN 18205, 1996). The latter section of the ISO specification stresses the need for decisions concerning the corporation's future mission. Paragraph B.6.1, 'Purpose', focuses on the profile of the company, its strategic aims, priorities, image and new areas of activity. Paragraph B.6.2 addresses the size of the organization, relative to other, similar enterprises, as well as market share and turnover and the number of employees. Paragraph B.6.3 deals with contextual aspects: national or local trends, such as social, commercial and technological and the availability of resources. And finally, paragraph B.6.4 deals with future changes within the corporation, planned expansions or reductions, and the reasons for these changes. All these considerations are relevant to the process of strategic planning. Based on them, decisions are made and documented in the strategic plan.

3.3   Who is involved in strategic planning?

Strategic planning takes place primarily at the corporate level of an organization. Managers adopt a long-term perspective to make decisions likely to affect the success of their business. A special strategy may be developed in a situation where a manager becomes aware that the mission and goals of the company are not being responded to. There can be different reasons for such situations arising, none of which may be related directly to buildings, either conceptually or in terms of a planned new building project or renovation. As developing such strategies may be outside the competencies of someone with an architectural background, other professionals with marketing, real estate, and financial skills may participate in the planning process, possibly assisted by project managers or facility managers.

Facility Management, recognized by corporate managers as an important management tool, includes the coordination of interrelated people, process and place issues within the corporation (see Chapter 8). Therefore, almost every strategy that is chosen to satisfy the organization's stated needs ultimately has some impact on business ('process'), buildings ('place') and the building users ('people'). A given strategy can offer several options regarding how these three elements can be combined, with processes maintained or changed, buildings reused or built new, and users involved or not.

User involvement is one of the major characteristics of building performance evaluation. In general, decision-making about buildings and space poses the risk of questioning, if not rejection, on the part of the organization's employees, if it is imposed as a top-down process. This may also be true during the strategic planning phase if decisions are made without the involvement of employees/building users. However, transparent decision-making processes usually involve building users and, if successful, respond to business needs, as well as user requirements. Therefore, the involvement of the knowledgeable user in strategic planning is crucial. It is an effective way to get a matching fit between the corporation and its accommodation or facilities. Moreover, for a corporation, such participatory processes are a credible expression of an organization's open culture and its humane values; while for the users, it is a big step towards increased empowerment, starting at this first phase of the building's life cycle. Ultimately, a successful planning process results in a good 'win-win' example that respects sustainability objectives and ensures ecological, economical and social benefits for all those involved.

3.4   Effectiveness review

As has been stated in the previous chapters, having an evaluative stance throughout the building delivery process and the entire building life cycle is a key characteristic of building performance evaluation.

Effectiveness review is the feedback loop applied to the strategic planning phase. It is a tool to review alternative strategies (options) and their attributes as they develop during strategic planning. Effectiveness review involves managers in focus interviews, as well as selected representatives from user groups in workshops of four to eight people (group interviews).

While departmental managers tend to evaluate possible strategies in terms of their specific business needs, user group representatives prefer to discuss how their basic needs and requirements are affected by attributes of the strategic plan. Both focused interviews and group workshops are organized and run by an external facilitator team. Being from outside the corporation, this team is considered to be unbiased in its role as facilitator and is respected by both sides: the managers with their corporate view and the employees with their view as process participants and prospective building users. The members of the facilitator team need to be good listeners and they need to have excellent communication skills. In order to stimulate the effectiveness review process, scenario planning may also be used, especially when possible options within the strategy need to be introduced (Brand, 1994). Input from interviewees and workshop participants is recorded on cards and displayed on the walls. As part of this transparent review process, the results of such meetings are summarized on charts and fed forward and presented to corporate managers. This 'bottom-up' planning process facilitates participation at all levels and ultimately guarantees a common understanding within the corporation of the strategy planned. Moreover, it gives managers at the corporate level the necessary feedback from user groups at the operational level.

Those elements of the strategic plan pertaining to the renovation, construction or leasing of space in buildings are reviewed in relation to departmental business needs, such as distance and time to market, staff availability, or big issue categories, such as corporate image or costs. These can then be used as performance criteria within the review process. Consequently, the performance of the different options within the strategic plan can be evaluated and compared against the above-mentioned criteria, and then rated with regard to their effectiveness. As a result of this review process, it becomes clear which option promises to be most effective. Finally, after passing through effectiveness review, the strategic plan becomes the basis for further action, for example, the beginning of programming. In this case, the preferred option favours a building project, the process having helped it to grow from a vague idea to a distinct and well-defined project.

3.5   Case study example

The following case study of a Swiss pharmaceutical company illustrates the process of effectiveness review during strategic planning.

3.5.1   Background

The pharmaceutical company in question operates internationally. It has a branch in the UK, located near London. The mother company in Switzerland decided to close research and production in the UK, but to keep administration, development and marketing there. This made research and production facilities available while administration staff occupied buildings that were partly owned, partly leased, and scattered around the production site. In addition, a site owned by the company was also available for new construction at that location. The Swiss strategy was to establish a UK corporate headquarters to support business needs in the UK, and to meet the basic needs of the staff, while respecting a sound economic rationale.

3.5.2   Procedure

Accordingly, several possible options first had to be identified that were consistent with the strategic plan of the mother company in Switzerland. These options included location at the regional or local levels, and building type (alteration of existing facilities, new structures, rent/lease of office space; see Figure 3.1). The process was coordinated by a facilitator team from Germany, situated – not only geographically – somewhere between the Swiss strategy and the UK requirements. The review process was mediated by this team.

Once a finite set of options had been identified, focus interviews with departmental managers were conducted in the UK to learn about their business needs. Every statement was recorded and visualized on cards, displayed on the walls and documented in the summary documents printed afterwards (Figure 3.2). Performance criteria were distilled from these needs, such as: closeness to market; recruitment of employees; wages; accessibility to airports; synergy effects of the clustering of like (pharmaceutical) facilities; proximity to authorities/agencies within the London area; and land and construction costs. These criteria were used to evaluate each of the previously established options.

Following an initial step of evaluation according to business needs, eleven options were left for further consideration, all located within the suburban 'green belt' area around London. Next, another series of group interviews was organized, in which representatives of user groups were invited to provide information about their basic requirements as users. Again, the information was recorded on cards as part of the transparent process of reviewing the effectiveness of the corporate strategy. Performance criteria were deduced from these requirements and, using Pena's methodology, were grouped into the categories 'function', 'form', 'cost', 'time' and 'context'. These criteria were applied to the evaluation of each of the eleven options. All options were rated on a 5-point scale according to these performance criteria, where '1' means low/bad and '5' means high/good (Figure 3.3).

		A	B	C	D	E	F
	Kind of building → Location of building	Change building into office space + add building	Renovate + add building	Rent/lease existing office space	Build new office space	Renew leases	Do nothing: downsize people
1	Main site	1A			1D		
2	Opposite site to East		2B		2D		
3	Adjacent site to North				3D		
4	New construction site			4C	4D		
5	Office building 1					5E	
6	Office building 2					6E	
7	< (within) 20 miles: Area 1			7C			
8	< (within) 20 miles: Area 2		8B				
9	> (out of) 20 miles			●	●		
10	elsewhere in UK			●	●		

Figure 3.1 Possible options.

Figure 3.2 Work process.

Figure 3.3 Option ratings.

Option — Ratings	Function (35%)									Form (30%)						Cost (15%)				Time (15%)			Context (5%)	Total
Criteria – Weight	Proximity to Railway Station (4)	Proximity to City Centre (3)	Availability of Parking (3)	Potential to facilitate Concept of Consolidation (5)	Potential to be Tailored for Specific User Requirements (4)	Potential to Facilitate Best Informal Communication (4)	Potential to Realize Interaction Needs/ Efficient Work Flow (5)	Potential for Good Work Environment (AC/Raised Floors) (4)	Complies with BCO Standards for Office Space (4)	Supporting New Image (4)	Potential to Attract High Quality People (5)	Potential for Landscape Design (2)	Potential of Flexibility to Change (5)	Potential to Grow (3)	Potential to Realize Non-Visible Security Measures (4)	Site-related Cost (3)	Building Cost–Efficiency (4)	Security of Cost (5)	Potential to Minimize Running Cost (4)	Occupancy within 3 years (5)	Occupancy within 4 years (5)	Security of Time (4)	Disturbance of Work Operations (3)	
1A Change Building on Main Site into Office Space	5	5	3	1	1	1	1	3	3	1	3	5	1	3	3	5	5	3	3	1	5	3	3	
1D Build New Building on Main Site	5	5	5	5	5	5	5	3	5	3	5	5	5	5	3	3	3	3	5	1	5	3	3	
2B Renovate + Add Building on Opposite Site to East	5	5	5	5	5	3	5	3	3	3	3	3	5	1	3	3	3	3	3	3	5	3	1	
2D Build New Building on Opposite Site to East	5	5	1	5	5	5	5	5	5	3	5	3	5	1	3	1	1	5	5	3	5	5	1	
3D Build New Building on Adjacent Site to North	5	5	3	5	3	3	3	5	5	3	3	3	5	5	3	5	5	1	5	5	3	5	5	
4D Build New Building on New Construction Site	3	3	3	5	5	5	5	5	5	5	5	1	5	5	5	5	5	5	3	5	5	5	5	
4C Lease New Office Space next to Construction Site	3	3	3	1	3	1	1	3	3	1	1	3	1	1	3	5	5	5	3	5	3	5	5	
5E Prolongation of Lease for Office Building 1	5	5	3	5	1	1	1	5	5	5	1	1	1	1	3	5	5	5	3	5	5	3	5	
6E Prolongation of Lease for Office Building 2	3	3	3	1	1	1	1	3	3	1	1	3	1	1	3	5	3	5	3	3	3	3	5	
7C Lease Office Building in Area 1	3	3	1	3	3	3	3	5	5	3	1	1	3	3	1	5	5	5	3	3	3	5	5	
8B Renovate + Add Office Building in Area 2	3	3	1	5	1	1	1	3	3	3	3	1	3	1	1	5	5	5	3	3	5	3	5	

Option – Results	Criteria - Weight																								Total
	Function 35%									**Form** 30%						**Cost** 15%				**Time** 15%			**Context** 5%	
	Proximity to Railway Station (4)	Proximity to City Centre (3)	Availability of Parking (3)	Potential to Facilitate Concept of Consolidation (5)	Potential to be Tailored for Specific User Requirements (4)	Potential to Facilitate Best Informal Communication (4)	Potential to Realize Efficient Work Flow Interaction Needs (5)	Potential for Good Work Environment (AC/Raised Floors) (4)	Complies with BCO Standards for Office Space (4)	Supporting New Image (4)	Potential to Attract High Quality People (5)	Potential for Landscape Design (2)	Potential of Flexibility to Change (5)	Potential to Grow (3)	Potential to Realize Non-Visible Security Measures (4)	Site-related Cost (3)	Building Cost-Efficiency (4)	Security of Cost (5)	Potential to Minimize Running Cost (4)	Occupancy within 3 years (5)	Occupancy within 4 years (5)	Security of Time (4)	Disturbance of Work Operations (3)	
1A Change Building on Main Site into Office Space	7,00	5,25	3,15	1,75	1,40	1,40	1,75	4,20	4,20	1,20	4,50	3,00	1,50	2,70	3,60	0,45	3,00	2,25	1,80	0,75	3,75	1,80	0,45	60,85
1D Build New Building on Main Site	7,00	5,25	5,25	8,75	7,00	7,00	8,75	7,00	7,00	3,60	7,50	3,00	7,50	2,70	3,60	0,45	1,80	2,25	3,00	0,75	3,75	1,80	0,45	105,15
2B Renovate + Add Building on Opposite Site to East	7,00	5,25	3,15	5,25	4,20	4,20	5,25	4,20	4,20	1,20	4,50	0,60	4,50	0,90	3,60	1,35	1,80	3,75	1,80	2,25	3,75	1,80	0,15	74,65
2D Build New Building on Opposite Site to East	7,00	5,25	3,15	8,75	7,00	7,00	8,75	7,00	7,00	3,60	7,50	0,60	7,50	0,90	3,60	1,35	0,60	3,75	3,00	2,25	3,75	3,00	0,15	102,45
3D Build New Building on Adjacent Site to North	7,00	5,25	1,05	5,25	4,20	4,20	5,25	7,00	7,00	1,20	4,50	0,60	1,50	0,90	1,20	0,45	0,60	0,75	3,00	2,25	2,25	1,80	0,75	67,95
4D Build New Building on New Construction Site	4,20	3,15	3,15	8,75	7,00	7,00	8,75	7,00	7,00	6,00	7,50	3,00	7,50	4,50	6,00	2,25	3,00	3,75	3,00	3,75	3,75	3,00	0,75	113,75
4C Lease New Office Space next to Construction Site	4,20	3,15	3,15	1,75	1,40	1,40	1,75	7,00	7,00	1,20	4,50	1,80	1,50	2,70	3,60	2,25	1,80	3,75	1,80	2,25	2,25	1,80	0,75	65,55
5E Prolongation of Lease for Office Building 1	7,00	5,25	1,05	1,75	1,40	1,40	1,75	4,20	4,20	1,20	1,50	0,60	1,50	0,90	3,60	2,25	1,80	3,75	1,80	3,75	3,75	3,00	0,75	58,15
6E Prolongation of Lease for Office Building 2	4,20	3,15	3,15	1,75	1,40	1,40	1,75	4,20	1,40	1,20	1,50	0,60	4,50	0,90	3,60	2,25	1,80	3,75	1,80	3,75	3,75	3,00	0,75	55,55
7C Lease Office Building in Area 1	4,20	3,15	1,05	5,25	1,40	1,40	1,75	7,00	7,00	3,60	4,50	0,60	4,50	2,70	3,60	2,25	1,80	3,75	1,80	2,25	2,25	3,00	0,75	69,55
8B Renovate + Add Office Building in Area 2	4,20	3,15	1,05	8,75	1,40	1,40	1,75	7,00	4,20	1,20	4,50	0,60	1,50	0,90	3,60	0,45	1,80	3,75	1,80	2,25	3,75	1,80	0,75	58,75

Figure 3.4 Option results.

3.5.3   Results

Finally, three of the eleven possible options turned out to be the most viable, regardless of the weighting of performance categories and individual criteria. While 'new structure' proved to be the best option compared to other kinds of buildings, the rating of the different locations showed a preference for sites within the local area (Figure 3.4).

3.5.4   Outcomes

All three options proved to be effective in terms of the Swiss company's strategic plan. As a result of involving real users, a wide range of possible options for the UK corporate

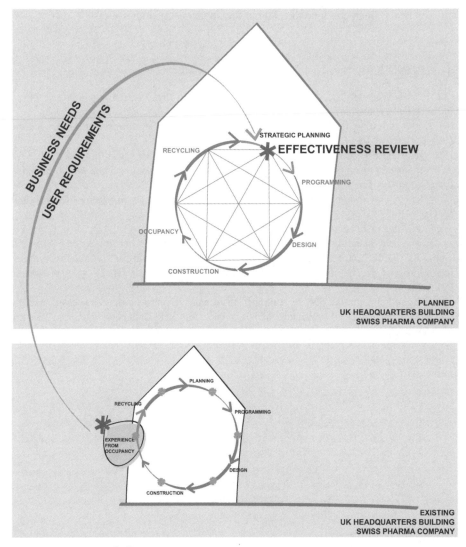

Figure 3.5 Importance of effectiveness review.

headquarters was narrowed down to the three that were most viable. Thus, from the perspective of the project in which this process would culminate, users had an early opportunity to provide input on the performance of their future building; and from the perspective of the organization's culture and values, they had a say in the process of decision-making and strategic planning. Following the reviewed strategic plan, facility programming was started for a new UK Corporate Headquarters building. Here, user requirements were analysed in more detail according to a variety of criteria. This next step involved more employees in participating in programming and, finally, led to a programme that respected their needs and requirements (Figure 3.5).

This example demonstrates how effectiveness review is an important evaluation loop within strategic planning. By using performance criteria to evaluate the strategic plan, issues of quality improvement are already being addressed, and the continuation of this process with its accompanying feedback loops into detailed programming helps to ensure quality considerations continue throughout the decision-making process.

References

Blythe, A. and Worthington, J. (2001). *Managing the Brief for Better Design*. Spon Press.

Brand, S. (1994). *How Buildings Learn: What Happens After They're Built*. Penguin Books.

DIN (Deutsches Institut für Normung) (1996). *Bedarfsplanung im Bauwesen (Brief for Building Design)* (DIN 18205). Beuth Verlag.

Hershberger, R. (1999). *Architectural Programming and Predesign Manager*. McGraw-Hill.

ISO (International Organization for Standardization) (1994). *Performance Standards in Building – Checklist for Briefing – Contents of brief for building design* (ISO 9699).

Kumlin, R. (1995). *Architectural Programming: Creative Techniques for Design Professionals*. McGraw-Hill.

Merriam-Webster's Collegiate Dictionary, 11th edition (2003).

Pena, W., Parshall, S. and Kelly, K. (1987). *Problem Seeking – an Architectural Primer*. AIA press.

Petronis, J. (1993). Strategic Asset Management: an Expanded Role for Facility Programmers. In *Professional Practice in Facility Programming* (W. Preiser, ed.), pp. 23–45. Van Nostrand Reinhold.

Preiser, W. and Schramm, U. (1997). Building Performance Evaluation. In *Time-Saver Standards for Architectural Data* (D. Watson, M.J. Crosbie, and J.H. Callender, eds), pp. 233–238. McGraw-Hill.

4

Phase 2: Programming/briefing – programme review

Alexi Marmot, Joanna Eley, and Stephen Bradley

Editorial comment

Programming is the second phase of building performance evaluation, and is directly linked to strategic planning. Known as briefing in the UK, programming can be approached in numerous different ways. This chapter reviews several methods in common use for engaging users in programming/briefing as the vital basis for design, including objective, subjective and collaborative techniques. The process of engaging users and moving towards consensus can often be as important as the documented programme itself.

Good buildings can be created only if we learn from existing buildings, both good and bad. In this chapter, the authors review methods of evidence-based common programming/briefing problems. The programme or brief has several levels of specificity: a vision statement; strategic brief; detailed brief; specialist and operational brief. Changes in the methods of procuring and operating buildings are affecting the timing of the brief and parties in charge of creating it.

4.1 Introduction

4.1.1 What is briefing?

The task of programming (North American usage) or briefing (UK usage) is to state what the building project is to achieve, in aspiration as well as function. Once completed, the brief serves as a set of instructions for the design and construction of the building. It may be for a new building, or for the refurbishment or refit of an existing property. The process of defining the brief is an important stage when priorities must be decided and consensus

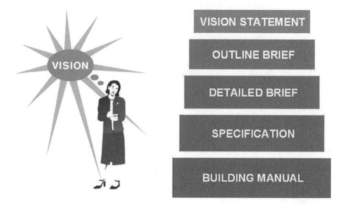

Figure 4.1 The stages of a brief.
(*Source*: CABE Commission for Architecture and the Built Environment (2003). *Creating Excellent Buildings: a Guide for Clients.* London: CABE)

reached, so that it results in clear and unambiguous information for the designers. The briefing stage takes time, especially when there are many groups whose needs and views need to be understood and taken into account. It must be carefully managed to reach agreement between all concerned parties and resolve competing needs.

Terminology describing the programme or brief is not consistent. Terms vary across the Atlantic, between different professions and different building types. Common terms include: outline brief; strategic brief; statement of goals; statement of needs (SON); statement of requirements (SOR); schematic brief – all used at early stages of defining the project – and brief; detailed brief; project brief; output brief or output specification; specialist brief; functional brief; operational brief, all used during project development.

In this chapter, the term 'brief' rather than 'programme' is used. Operational briefs, which can be transformed into maintenance and management information, are not covered except as the requirements for operating a building, which are reflected in the strategic and detailed brief (see Chapter 6 on commissioning). In small projects, only a strategic brief is needed. Large, complex projects need all briefing stages to be considered thoroughly.

4.1.2   Briefing in the BPE framework

Within the framework of building performance evaluation (BPE), briefing (programming), is positioned after planning and before design. In reality, some early briefing occurs in the planning stage, and much of the detailed briefing occurs in the design stage (see Figure 4.2). The detailed brief for some specialist elements may even await the start of construction. Feed-forward from a post-occupancy evaluation is also a form of briefing, as explained in Chapter 1 of this book.

Ideally, the requirements encapsulated in each of these phases are subject to formal review and sign-off prior to proceeding, and each succeeding phase starts with a full team review of the project vision, and the strategic and detailed brief. Without such a review process, an effective transition from strategic plan objectives to detailed design is at best achieved through solid custodianship from one or more team members but, at worst, can result in complete discontinuity.

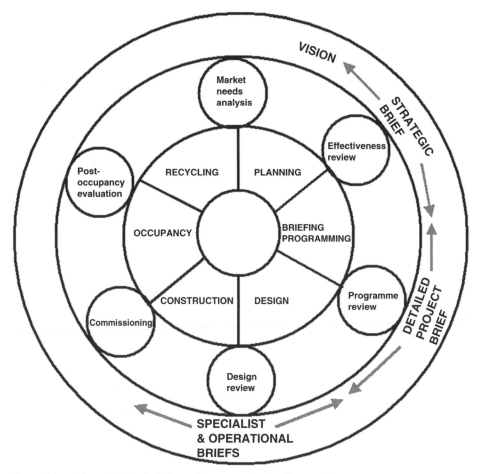

Figure 4.2 Briefing within the building performance evaluation framework.

An everyday analogy for the use of the above terms is illustrated by a personal decision to buy a new mobile phone. The information needed by the purchaser is described in Figure 4.3, under the briefing stages identified earlier. The example of the brief for a mobile telephone contains several messages that have parallels within building projects. The clear statement of the strategic brief should always be respected whatever the particularities of the more detailed requirements. The issues of usability and cost in use are very important and must be balanced with other criteria such as functionality and style.

4.2   How are briefs prepared?

Four approaches to preparing a brief for a building project are explored in this chapter. Their use depends on the complexity and scale of the building project, and the experience of the client. In each approach, different people ('briefing agents') are responsible for

Briefing the acquisition of a mobile phone or PDA			
Strategic	Detailed (functional)	Technical	Operational
Small, light, easy to use, stylish	Balance of functionality between voice, text, data, camera/video and games; length of battery life, time to recharge and weight of the recharger; screen size and colour; size and nature of keypad; network band coverage; barring possibilities; payment terms	Wireless connection with other devices; data download speed	Itemized billing; replacement of lost or stolen phone

Figure 4.3 Briefing stages – an analogy.

preparing the brief. These include:

– the client directly;
– a specialist brief-writer;
– the architect as part of the design process;
– people in the teams who will finance, deliver and operate the building.

The approach taken is contingent on the complexity of the building project, the relationship between the client and the rest of the project delivery team, and on whether the client is the owner, occupier and/or manager of the finished building. A single body acting as the commissioning client, owner, occupier or manager of the building allows a straightforward, direct briefing process. When those with an interest in long-term involvement in the building (users, managers) are split from the commissioning client, additional considerations arise. In the UK, many current new construction projects separate the user from the organization that designs, builds, finances and operates the building. This is referred to as the Design, Build, Finance and Operate (DBFO) process, prevalent in large public sector projects such as schools, universities, hospitals, subsidized housing and administrative buildings. In such projects, the commissioning client gives an outline brief to the DBFO organization, which turns it into a detailed brief.

Typical projects best suited to the different 'briefing agents' are described below.

4.2.1   Brief prepared by the client directly

This is best suited to small projects such as a small office, small retail outlet or a private home, commissioned by a small company, a single individual or a family. These are generally small simple buildings, with straightforward functional requirements and a limited group of users whose needs are readily understood and documented. This approach is only suitable for large projects for serial clients with an experienced internal property team and an agreed set of design and accommodation standards; for example, the creation of new outlets for a multi-locational retail chain or High Street bank. In the public sector, this approach is common for repeat projects such as children's nurseries, or small court buildings. In these cases, a formal brief is handed over to the design team, often in the form of a large manual specifying every element of the building.

The benefit of clients directly preparing the brief is that they have the most direct understanding of their real needs, even if occasionally they lack the technical expertise to convey detailed constructional or operational requirements.

4.2.2   Brief prepared by an architect

A common approach to briefing is for the architect to develop the brief as part of the design process. This is enshrined in the contract between architect and client in the UK in the Royal Institute of British Architects Plan of Work (RIBA, 1998). By asking the client and different stakeholders, such as users, visitors, local neighbours, planning authorities, and interest groups what they need or would find unacceptable, the architect forms a comprehensive understanding of what should be designed. If needs conflict, then the architect seeks to find a design solution that satisfies the different parties. For example, if the people who will use the building demand a great deal of accommodation and car parking, while the local planning authority and neighbours prefer a small building with little parking, then the architect typically will seek a resolution; for example, by mitigating the impact of a building on its neighbours by means of 'environmentally-friendly' design and provision of high quality landscaping.

Small architectural firms typically conduct briefing informally, by asking questions and occasionally observing the client in action. This can build understanding and trust between both parties – the architect becomes familiar with the client and the client learns to trust the architect and feel that their needs are being heeded.

In larger firms, people other than the project designers may ask questions of the client, therefore the designer is less able to develop an intuitive understanding. More formal methods of recording needs are applied, such as an accommodation schedule listing each space and required floor area and functional requirements, thereby providing specific, though limited, information in a structured way for use by all design team members.

The main potential benefit of the architect as brief-writer is the close proximity of the designer to the client's needs and knowledge. The main disadvantage, when the designer is also the brief-writer, is that there is no expert on the client's side to catch and correct situations when, for example, the design solution fails to reconcile all the competing needs, or some needs are ignored. Clients may assume that all their needs have been met because they have been consulted, and they may not readily comprehend technical drawings. Another shortcoming can be that the new project simply emulates what has gone before, albeit in new 'clothing'. While all these risks apply to every project, they are worsened if neither the client nor the architect has special briefing expertise, and if no independent person makes a critique of the evolving scheme.

4.2.3   Brief prepared by a specialist brief-writer

Brief-writing is a small, but growing specialty within building design. It may be conducted by a specialist team within a large multi-disciplinary architectural firm or by a smaller, independent expert firm. It has arisen alongside the growth of other specializations, such as ecological and environmental design, historic building conservation and urban design. Reasons for the growth of briefing specialists include: the increase in project size and complexity; more extensive technical requirements; and the internal organizational complexity of large clients. To understand the needs of the many different stakeholders and to keep track of changes across the extended timeline of a large construction project, requires expertise in facilitating meetings, sometimes controversial, with patience, persistence and persuasiveness, as well as encouraging the flow of information.

4.2.4　Brief prepared by several parties in new procurement arrangements

Under procurement systems, such as Design and Build (D&B), or Design, Build, Finance and Operate (DBFO), several parties need to participate in clarifying the brief. Clients need to define the strategic brief and to specify (in as much detail as possible) their detailed needs for the building and its operation. Lawyers for the client normally ensure that the brief is included in the terms of the contract signed between the clients and the company that will provide (and possibly operate) the building. The challenge for the client is, therefore, to include in the brief as much detail as they can about imagined future needs, right at the start of the procurement process before designers are appointed.

Meanwhile, the D&B or DBFO team and their designers may prepare a brief that includes their own interpretation of needs from their perspective of buildability and ease of operation. The client needs to check that the contractor's own brief and design solution and the resulting building meet the original client-side requirements.

4.3　Common briefing problems

Industry experience in the UK and North America indicates that systematic programming/briefing and review processes are far from widespread and often are not taken beyond lip-service. Case study research undertaken in the UK by Barrett and Stanley (1999) shows that there is 'no simple "cookbook" solution to good briefing but there are clear areas where improvement effort can be productively invested'. In the introduction to David Hyams' UK construction guide to briefing, Chappell notes the lack of a typical, widely applied set of processes, and comments on the attitude of architects to the relationship of briefing and design: '... there appears to be very little order in the way in which architects approach the task of securing the brief ... Many architects treat it as a casual endeavour, an almost unnecessary precursor to the design stage.' (Hyams, 2001). In this view, design, rather than briefing, is seen as the way to define and negotiate the scope and qualitative goals of the project. It does not allow for a thorough understanding of users' needs and systematic feedback to design decision-making.

In discussing the relationship of programming to design from the US perspective, Cherry (1999, pp. 12–13) assumes that 'the architect should be skilled at all stages of architectural programming and should be able to provide the service'. However, Vischer (1996), also writing from a North American perspective, comments that 'the period of information collection about how people will use the space to be designed ... is sometimes cursory, and often entrusted by naïve clients to design professionals with little experience or interest in pre-design programming'. In this scenario, briefing is limited to projections of the number, type and activities of the future users of the space to be provided. This sort of process, even if based on user surveys providing some qualitative information about user preferences, does not engage stakeholders in informed debate about the performance of the building they are to occupy, still less involve them in decision-making and ownership of solutions. In such circumstances, the value of building performance evaluation can be severely constrained and, more significantly, the opportunity is lost for improving the occupancy processes of users alongside building planning and design, and engaging users in a 'change-generating process'.

4.3.1   No brief

There is evidence from the Commission for Architecture and the Built Environment project-enabling programme in the UK suggesting that, when building projects go awry, it is often due to an inadequate or non-existent brief. This situation commonly occurs with first-time clients who are uncertain how best to embark upon a building project. They may be unclear about their needs except at the most general level. They may not know what a brief is, or that they need to have one. They may appoint an architect or a builder to design a new structure or to refit an existing one without ever fully stating their needs in terms of functionality, timetable, budget and appearance. The clients then run the risk of agreeing to a design that has evolved through a process of small, incremental decisions while omitting some of their more fundamental needs.

4.3.2   Incomplete or poorly timed information

Sometimes only partial information about users' needs is available at the time when building design is undertaken. Even the strategic requirement may be sketchy and may not help the main design decisions to be taken in an intelligent fashion. Naturally, a detailed brief can be developed later to cover the fitting out of the building, the selection of furniture and small elements. An example is a recent new building for the new Greater London Authority office. This highly specific building was commissioned even before there were any permanent members of staff. The functions and size of the building were decided on the basis of informed guesses about their requirements. Only when the new building was nearing completion was it possible to take a detailed brief from the newly elected council members and appointed officers. As a result, the new organization had to fit into a set of spaces that could only be given minor adjustments.

4.3.3   Unambitious briefing

Positive outcomes from many building projects arise from using building change as a stimulus to rethinking the organization and its activities. Organizational change that may have proved difficult to introduce otherwise can sometimes be introduced via a building project. Examples abound, such as streamlined processes for handling work flow in outpatient clinics introduced alongside a new hospital facility; or of a new office building encouraging more communication between teams through its interior design; or a new or refurbished retail store or hotel repositioning itself in its market. But these ambitious goals are not common. Often the brief is no more than a list of functional spaces. The resultant building is a missed opportunity for imaginative development and change.

4.3.4   Ignored briefing

Even when a clear and imaginative brief has been prepared, it is sometimes overlooked or ignored by the design team. Client concerns may fade into the background while the designers face the front-line battle to reconcile practical concerns of meeting deadlines and budget estimates, squeezing space into the site, negotiating with planning authorities and resident groups, and considering all the technical demands of foundations, structure, mechanical and electrical systems. Time, commercial pressures and a tendency to develop design solutions before full appreciation of broad performance needs and value have been taken on board, often compromise the recommended sequence of brief preparation and review preceding design that is described in the *Integrative Framework for Building Performance Evaluation* (Preiser and Schramm, 1997).

4.4   Techniques for briefing

Three categories of techniques are useful for collecting information to be used in the brief:

– objective
– subjective
– collaborative

The description below covers a range of techniques, including all those that form the AMA Workware Tool kit that has been used to gather information for briefing on more than 26000 people in 170 buildings. A description of this tool kit can be found at www.usablebuildings.co.uk.

4.4.1   Objective techniques

Audits: Space, cost, FM, energy, accessibility
Audits carried out on existing premises are an essential baseline. Knowing the starting point makes it easy to understand the current operation, provokes thought about potential improvements, and helps to define beneficial changes on the basis of evidence. Hard evidence allows comparison with industry benchmarks – for example, whether or not schoolchildren are being given too little space for sport compared with national averages or aspirations. Hard evidence also allows designs to be checked as they evolve. Clear definitions must be given for all objective measurements to enable comparison of like with like.

Space occupancy, space use over time
This documents what actually happens over time in built space – how many people use particular spaces for what activity. Observers gather the data using standardized

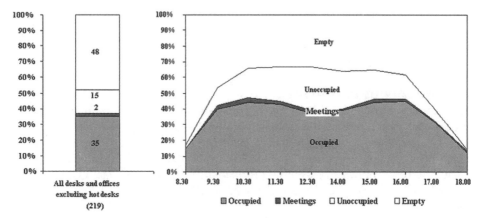

Figure 4.4 Typical data from studies of space use over time.

techniques pioneered by Alexi Marmot and Joanna Eley (Marmot and Eley, 2000), taking spot checks at agreed time intervals. Web-cams are sometimes a suitable supportive tool, for example, documenting flows of people through shopping malls. The data generated help to provide evidence for the purposes of sizing the new building.

Feed-forward from post-occupancy evaluations
Gathering and documenting best-practice lessons from post-occupancy evaluations is essential to better briefing. Chapter 7 elaborates on post-occupancy evaluation in the context of building performance evaluation, and examples of POE in a variety of different contexts are found in Chapters 11, 12 and 13 of this book.

The techniques referred to above are suitable for buildings of all sizes and are used to help plan the necessary changes to working patterns resulting from moving into a new building, as well as to determine how to design the space.

Performance specifications
This technique sets targets for delivery of specific performance, such as temperature, humidity, durability of materials, average waiting time for lifts (elevators), etc. Specifications are commonly used to define engineering outputs. A building design can be tested while it is evolving to ensure the specification is met. It can also be used to set out, in measurable terms, desired outcomes of investment in building. As shown in Chapters 1 and 2, performance criteria have an important role in BPE.

4.4.2 Subjective techniques

Interviews
Interviews can be short and sharp, or more lengthy. Selecting who is to be interviewed needs to be done carefully. For strategic decisions and predicting the future needs of the organization, only the most senior people need to be interviewed. For detailed

understanding of the number, roles and needs of people in local groups, it is essential to interview people at other levels. Functional managers in Human Resources, Facilities Management and Information and Communications Technology should also be included.

Questionnaires and diaries

Asking future users and stakeholders how they use their current facilities and what changes they most desire provides subjective information to incorporate into the brief. At the same time as giving 'soft' data about what is needed, it also gives a shared ownership of the brief, helping to promote commitment to the future project (see Section 4.4.3 below). Surveys can be carried out by distributing and collecting paper questionnaires, electronically via email and web page, or in person, with structured questioning from an interviewer.

Observations by design team

When a design team can spend time with their client in the current building, they should systematically note aspects for the future – positive aspects to be emulated and negative characteristics to be avoided. If there is no current building, then visits to similar facilities should be made for this purpose. Photographs and videos are also useful tools.

4.4.3   Collaborative and communicative techniques

Visits and exploration of precedent

A common problem when a new building is to be planned is that users often have no idea of alternatives to their current environment. Expanding their horizons by visits to different buildings helps them to imagine their own brief in a new light. The added value of visits is the time spent together as a team developing shared understanding and goals. Case studies drawn from published material, videos and presentations are slightly less valuable but expand the range of what the briefing team can experience.

Workshops, focus groups, brainstorms, group walkabout

All of these techniques bring people together to debate and agree on the brief for their project. At early project stages the material considered is general; as the project progresses focus is often needed on specific topics, such as security, catering, or storage. Facilitated value and risk management workshops help refine the brief and priorities. Techniques that may be used include issues sheets, analysis cards, gaming cards, images and precedents, and mood boards. Open public meetings may sometimes need to be held with stakeholders to inform them about the project and to learn their requirements.

Room data sheets

For detailed design, it is essential that a full brief is created for each and every space in the building. Room data sheets need to describe the requirements of each space in the new accommodation. They should state the intended use and number of users, hours of opening, size, daylight and orientation, building services (heating, lighting, air handling, power, telecoms, water and drainage, etc.), finishes, furniture and equipment. Room data sheets need input from current users, as well as the project team and designers, before they can be signed off. It is essential that predictions about future changes in the space and its uses are also considered; for example, what will be the impact of changes in working patterns and nanotechnology on furniture and equipment?

4.5   Contextual issues for consideration

4.5.1   A knowledge-based industry

Building design, construction, and management are becoming more reflective and knowledgeable about what constitutes good practice. Reviews of large projects are now mandatory in the public sector in most western countries. Private sector organizations also frequently carry out post-project and post-occupancy evaluation for internal and external audit. The accumulation of evidence of good practice from these reviews can and should be incorporated into future briefs. As the evaluation of building performance becomes more common, and as public domain knowledge increases, it should become easier to prepare better briefs, and for them to be accepted and acted upon by designers.

4.5.2   Closing the loop from concept to the building-in-use

The UK has pioneered the creation of integrated supply chains for construction, where all those concerned with creating the building, from architect and consulting engineer to construction manager, trade contractors and specialist materials and systems suppliers, are involved early in the process. They work together to ensure that their values and assumptions are made explicit and that their skills and knowledge contribute to the whole process. While still in relative infancy, the ideal version of an integrated process has one party responsible for intelligent briefing, design, construction, operation and maintenance, who may specialize in group process facilitation. This approach gives an opportunity to apply a long-term perspective to aligning and optimizing operational requirements with functional and financing requirements. The BPE framework could usefully be applied to this approach.

4.5.3   Timing

A fundamental problem in briefing for large projects is timing. Designers need large quantities of consistent and reliable information right from the start of their process. For new-build projects, information is often required several years before the project is ready for use and should look ahead for many years, even decades, when the building will still be in use. Yet people within client and stakeholder organizations are often reluctant to take the time to think about future needs. They focus on immediate deadlines or immediate problems that need to be solved. So the brief, as received from users, must be tempered and reinterpreted by the brief-writer so that it is meaningful to the time-horizons of the designer, and enables the possibilities of future-proofing the building to be explored and evaluated.

4.5.4   Rapid information transfer

Better systems for gathering and exchanging information can improve briefing. One example is electronic voting and electronic suggestions collection for all users to input their needs for rapid consideration by the brief-writers. Another example is web-based project information systems, enabling a variety of applications and interested parties to have access to and comment on an integrated, 'live' set of databases of current information on project requirements and designs.

4.5.5   Client sign-off

A worthwhile brief not only reflects the most up-to-date knowledge of good practice in buildings and what users want, but is also fully accepted and adopted by the client and fulfilled by designers. Structured sign-off by the client is essential at several stages as the project progresses from strategy to details. If client needs change after each sign-off, the impact on the design, budget and project timetable can be separately identified. This allows the client the opportunity to make value judgements as to whether the changes are mandatory for project success, or if they can be incorporated as minor alterations after project completion.

At every stage of the project, the brief must be able to provide a necessary and sufficient level of information for the designers. It must convey information about what is needed and why, so that the design team can balance these needs against what it is possible to do with the site, the budget and the timetable. Communications media may include written

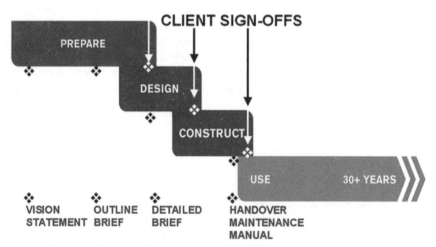

Figure 4.5 Client sign-off of the brief.
(*Source*: CABE Commission for Architecture and the Built Environment (2003). *Creating Excellent Buildings: a Guide for Clients.* London: CABE)

statements, lists of requirements, room data sheets, timetabling information, organizational charts, diagrams, and illustrations.

4.6   Conclusions

The best buildings are not only excellent exemplars of form, grace, aesthetics, and building technology, they also help the development of the organizations that use them. Well-briefed and well-designed office buildings reflect the values and aspirations of the occupying organization and help to communicate them to users and the wider public. Well-briefed and well-designed new hospitals help patients feel better and staff work efficiently and effectively in delivering health services. Well-briefed and well-designed new schools help pupils and staff feel valued and sustain new ways of teaching and learning. They can only do this if the brief is intelligently prepared, researched and clearly articulated, and effectively assimilated by the designers.

The ideal process of briefing engages both supply and demand sides and becomes a collaborative learning process. It builds on feed-forward from the strategic planning process and from evaluation of past projects and facilities in use. But it also involves the design team in understanding the client vision and technical requirement, so that design creativity can help to invent new and better solutions to organizational goals.

References

These include publications cited in the text and several other publications of note on briefing.

Barrett, P. and Stanley, C. (1999). *Better Construction Briefing*. Blackwell Scientific.

Blyth, A. and Worthington, J. (2001). *Managing the Brief for Better Design*. Spon.

Bradley, S. (2002). What's working: briefing and evaluating workplace performance improvement. *Journal of Corporate Real Estate*, Vol. 4, **2**. Henry Stewart Publications.

CABE, Commission for Architecture and the Built Environment (2003). *Creating Excellent Buildings: a Guide for Clients*. London: CABE.

Cherry, E. (1999). *Programming for Design: from Theory to Practice*. Wiley.

Horgen, T.H., Joroff, M.L., Porter, W.L. and Schon, D.A. (1998). *Excellence by Design: Transforming Workplace and Work Practice*. John Wiley.

Hyams, D. (2001). *Construction Companion to Briefing*. RIBA Publications.

Marmot, A.F. and Eley, J. (2000). *Office Space Planning: Designing for Tomorrow's Workplace*. McGraw-Hill.

Pena, W., Parshall, S. and Kelly, K. (1987). *Problem Seeking: an Architectural Programming Primer*. Washington, DC: AIA Press.

Preiser, W.F.E. and Schramm, U. (1997). Building performance evaluation. In *Time-Saver Standards for Architectural Design Data* (D. Watson et al., eds), pp. 231–238. McGraw-Hill.

Royal Institute of British Architects (1998). *The Architect's Plan of Work*. RIBA Publications.

Salisbury, F. (1998). *Briefing Your Architect*. London: Architectural Press.

Vischer, J.C. (1996). *Workspace Strategies: Environment as a Tool for Work*. Chapman & Hall.

5

Phase 3: Design – design review

Jacqueline C. Vischer

Editorial comment

Design review occurs in the third phase of building performance evaluation (BPE). After strategic planning, followed by programming (briefing), participants become involved in the actual process of creating spatial and physical solutions for clients. Design, and the evaluative loop of design review, is the phase when users and decision-makers begin to see what kinds of spatial solutions are available and appropriate. In determining options that fit into existing constraints and solve problems effectively, designers develop two- and three-dimensional images that begin to respond to the priorities established throughout the strategic planning and functional programming processes (Preiser and Schramm, 1997).

Designing and reviewing designs is a process that can take many forms. The design team produces ideas and the graphic representations to communicate them. Depending on the circumstances of the project, these can be reviewed by clients, users, an outside body (for example, a design review board), as well as by the technical and financial consultants, and by the builder or contractor. Although feedback and input from these stakeholders is valuable and important, the major difficulty in any design review process is ensuring that the design team retains the integrity and quality of its ideas, while at the same time being responsive to the changes others suggest, impose or insist upon.

5.1   Introduction: defining design review

Design review is a critical stage in the BPE process because, at the outset, the design team and the client/users are on different sides of the table. The designers are generating and giving form to ideas; the client and users are reacting, reviewing and providing input, but not necessarily providing ideas. In conventional situations, after the intimate and dynamic phase of programming (fully described in Chapter 4), the design team separates somewhat from the client-user side. The designers go off to work on their own, to exercise their skills, to synthesize the information they have received into solutions to the problems they have identified. However, in building performance evaluation, the process requires ongoing evaluation and feedback, therefore the design review loop mitigates against this separateness in the creative process and keeps everyone working together.

The BPE approach provides a process and procedures that enable designers to work productively while at the same time staying connected to the people they are designing for (see Chapters 1 and 2). Rather than going away and coming back with some visually attractive drawings that clients and users then react to, as in a conventional process, in BPE the designers make their process and thinking explicit, show how the various constraints on problem solution are shaping their thinking, and retain authority throughout the process. This process must be carefully managed (even facilitated, as explained in Chapter 18) to ensure that good design ideas are not lost through miscommunication, stakeholders' fixed ideas, or new information becoming available at a late stage and changing the definition of the problem.

Design review may also occur outside the immediate user-designer context, and take the form of a formal evaluation by an outside body with authority over design decisions but without a role in the planning and programming process. For example, many North American cities have design review boards who meet to evaluate major new architectural projects, purely according to criteria related to form and appearance, contextual appropriateness, choice of materials, accessibility and functional adequacy (Scheer and Preiser, 1994). Current public debate about the buildings being designed to replace the two lost towers of New York's World Trade Center provide a case in point.

If strategic planning, as described in Chapter 3, and functional programming (briefing), as outlined in Chapter 4, have not been performed successfully or completely, the flaws in decision-making and the questions to which answers have not been found will show up in design review. In fact, even where the first two phases of BPE have been effectively and responsibly carried out, the simple transformation process from words to images, as plans take shape on paper and sketches show perspectives and elevations, is almost sure to call up new, unsuspected questions that no one thought of previously. This is an appropriate role for design review and one that should be anticipated by designers and clients. However, if the questions are too numerous, or take time to answer, or somehow slow down the process because no one wants to deal with them, then the design review stage is taking on a role it should not have – compensating for inadequate programming. As a result, this stage can become prolonged, increasing the cost to the designers, if they are working on a fixed fee contract, and/or to the client, if they are paying for changes to the design.

In summary, the design review stage is best thought of as that phase in the BPE process where, together – but occupying different roles – designers, clients and users decide on form. The spaces, the connections between spaces, the sizes, shapes and quality of the spaces will emerge from this stage. Building team members will draw on goals and priorities set in the strategic planning process, on information about use and occupancy contained in the programme, and on additional information such as cost estimates, colours and materials, as well as on outside influences such as zoning regulations, building code requirements and likely future use and adaptation over the lifetime of the building, as they move through decisions about building form. These decisions include critical issues of cost and quality, and the trade-offs that have to be made (see Chapter 19). As others have pointed out, every design decision is a balancing act between time saved, costs reduced and maximizing quality, with most decisions only responding to two out of these three criteria.

5.2   Implementing design review

Who participates in this process? This varies depending on the building project. Generally, the larger and more complex the building, the greater the number of groups likely to be affected and, therefore, to want a say in its design. Complex institutions, such as prisons and hospitals, have numerous technical and security requirements, thus increasing the urgency of adequate design review at an appropriately early stage.

In simpler projects, or those consisting of renovations to an existing building, the number of interested parties may be fewer. Hopefully, the programming stage has enabled major conflicts, needs differences, and varying priorities to be resolved, such that interest groups do not form to pressure for their own solutions after the design has begun to take shape. If there is not consensus among user groups by the time planning and programming are completed, the transformation of text into images during design review is a process that guarantees to bring out and make known unresolved differences in users' viewpoints (Vischer, 1999).

In addition to the information contained in the final programme, there is no ironclad rule against using design images and sketches to generate discussion and encourage negotiation among holders of differing opinions. However, some occasions are more propitious than others for such open discussion. In many projects, clients are too concerned about the cost of time passing, and/or not committed enough to broad-based user participation to allow design review to serve the purpose of bringing interest groups to consensus. So some stakeholders are excluded from design review, the client being satisfied that they were adequately consulted during programming, or they are invited to participate at a later stage when many key decisions have already been made.

5.3   Tools and skills for design review

One of the objectives of the design review stage is to enable designers to receive user feedback on the design early enough in the process so that correcting mistakes and making improvements is simple and cost-effective. In addition to techniques of getting stakeholders together and organizing a process of negotiation and exchange, a number of computer-based techniques are available to access knowledge and apply new knowledge directly to design. As described in Chapter 2, designers may draw on pre-existing design guides: generic principles derived from research into a specific building type and published to guide design decisions. They may also access published standards, ranging from the prescriptive standards of national building and fire codes to more qualitative, performance standards for ensuring comfort and quality for building users. Finally, designers may refer to existing examples of the same or similar building types, which have been held up as examples of effective design decisions, energy management, user functionality, etc. By using such tools as these, a designer is already using POE or user evaluation early in the design stages, instead of waiting until the building is built and occupied. Thus, designers are able to consider the effects of design decisions from various perspectives while it is still not too costly to make changes (Vischer, 1995).

One useful technique is to have a member of the team act as a 'facilitator' to ensure that all interest groups have a chance to express their concerns during the design process, and to monitor the process such that suggestions are made constructively and battle lines are not drawn. On the other hand, the architect or designer needs to be sure that people who are invited to express their viewpoints do not assume they have been given control of the process. One of the difficult aspects of design review is to ensure a balance between the professional knowledge of the designer and the mixture of knowledge, hopes, wishes and illusions that often characterizes other participants' perspectives. Thus a 'neutral' person playing the role of process facilitator – as described in Chapter 18 – can ensure that all points of view are represented, as well as ensuring that design ideas are clearly communicated and understood.

Other techniques include pinning up sketches and renderings of the project for users at all levels of decision-making to examine at their leisure. This might occur in an occupied building, to ensure that the ultimate users are involved; it may also occur at scheduled meetings. Who participates in these meetings and how they are run is critical to the quality of feedback received during design review. Computer tools such as full-scale mock-ups, simulations of interior and exterior spaces, and virtual walkthroughs can provide a checklist against which plans and drawings can be evaluated. Thus the designer receives information to correct technical and construction aspects of the design concept. In addition, face-to-face meetings in which stakeholders respond to the designers' architectural ideas and design development can sometimes uncover information not previously identified.

Design review can be an emotional and stormy time on any architectural project. However, if programming has been thorough, participatory, and comprehensive, the stage of design review will be shorter, more efficient and more comfortable for all participants.

5.4   Design review: a case study example

Design review was extensively used in the planning and designing of a new building (plant and offices) for a New Hampshire (USA) manufacturing company. One of the reasons for the new building was to accommodate a situation of radical organizational change while avoiding costly design changes and change orders during construction. Thus programming and design review were critical stages in the process, and design development of working drawings and the construction phase were relatively fast and painless. The company planned a major facilities expansion in response to increasing sales that had led to overcrowding and inefficient space-use in their existing facility. At the start of the project, a local architecture firm developed an architectural programme that contained information about adjacencies and square footage, as well as future growth, which could be used for estimating costs and for applying for building permits and zoning approval.

Subsequently, the CEO determined that this linear approach to programming and allocation of workspace did not serve his own ideals of employee empowerment, consensual decision-making and organizational change. A 'facilitator' was hired to ensure that the organizational change process was integrated with and indeed part of the design process, and that every employee in the company was informed, consulted and helped to buy in (Zeisel, forthcoming). The workspace design process that was developed included the following stages.

5.4.1

Creating a shared vision of the new space, setting objectives, defining the concept, and coming to consensus around the decision to implement it. This work was primarily carried

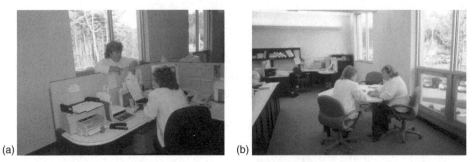

(a)   (b)

Figure 5.1 Hypertherm employees trying out mock-up workstations. Photo: J. Vischer.

out by members of the senior management team, who then took responsibility for communicating their decisions to team leaders and other employees, and answering employees' questions about the proposed changes.

5.4.2

Setting up a pilot study to test the new workspace ideas, and having an actual work team evaluate them in the Hypertherm organization. The construction of a simulated workspace showed not only alternative physical configurations, but also tested one of the new organizational groupings (multi-functional teams), allowing employees to familiarize themselves with what was being proposed, and to comment and provide feedback both to their own managers and to the design team (see Figure 5.1).

5.4.3

The stages of design review were critical in this project because programming by the architect had been limited to lists of spaces and square footage and had not addressed organizational change. In fact, it had not been possible to programme fully the new work environment until some of the management and organizational changes had been designed and implemented. Thus design review included efforts of both senior managers (including the CEO) and the design team to communicate the design concept for the new workspace to all employees, as well as various strategies and techniques to generate useful feedback from employees and, ultimately, to get their buy-in. These efforts included all-company meetings, special in-house publications to disseminate information, and many departmental work sessions where managers fielded questions from staff.

5.4.4

Once employees were clearly informed of the details of the restructured organization, spokespersons were selected for the new, multi-functional teams, to work with the design team and to participate in design review. There were a number of stages to this process. Hypertherm felt that unless time was spent both sharing the new workspace ideas with

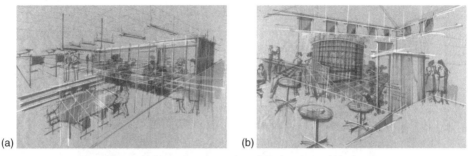

Figure 5.2 Sketches of the new workspace to share ideas about how it would look. Designer: Robert Charest.

Figure 5.3 Open workstations for senior management in the new building. Photo: J. Vischer.

employees, and getting feedback from them on how it would work, the advantages of the new work environment that were envisioned by the CEO and senior managers would not materialize.

5.4.5

Once design review with users was completed, it was necessary to continue the work with the general contractor and engineers. As construction got under way, technical and cost questions were raised and answers had to be found. The temptation was for the client, under pressure from the contractor to keep costs down and to make decisions quickly, to forget earlier decisions embracing innovation, and to accept traditional building solutions. For example, although everyone was in agreement regarding higher ceilings to enable indirect ceiling lighting, the higher slab-to-slab distance caused the contractor to increase his cost estimate, almost forcing the client to revert back to traditional ceiling heights that would have meant lighting the offices with conventional overhead lighting. In another example, the contractor found it hard to believe the CEO and senior managers would occupy the

Figure 5.4 Typical teamspace configuration in the new offices. Photo: J. Vischer.

same, open plan workstations as all the other employees, and privately tried to convince them to build five enclosed private offices instead.

5.4.6

Once the building was occupied, feedback from employees was sought to help make decisions about managing the new space, making changes and fine-tuning space needs, and educating new hires who had not gone through the design participation process. A post-occupancy evaluation was carried out in which employees were surveyed regarding their adaptation to the new work environment, and managers were interviewed to determine whether or not key business processes had improved as the result. The results were especially positive for those teams and team representatives who had participated the most in the process, and for whom, in fact, the changes were the most radical.

The main advantage of employee participation in design review, and the one that paid off most for the company, was the understanding and acceptance of team workspace and how it should work. Everyone in the company has 48 sq. ft. workstations with 52-inch partitions, including the CEO and senior management team. Everyone has two pedestal filing cabinets, and each team has shared filing cabinets. Each team also has a team worktable integrated into its teamspace, along with printers, faxes and other necessities. On each floor, a small number of enclosed rooms are available on an as-needed basis, for working alone, private meetings, and small group work sessions. Airy conference rooms were built throughout the facility for larger group meetings. While many of these decisions were controversial when they were taken, for example, the reduction in workstation size and the increase in shared workspace, all participants agreed that these design elements were necessary for effective teamwork.

All the spaces were immediately used at move-in, saving the company time both on revisions and redesign, as well as on dealing with employee resistance and resentment.

The survey results, compared to the results of the pre-design user survey, indicated that not only were respondents pleased with their new space, but also the process of buy-in and participation in design helped them anticipate how it would work. They accepted it without the period of discomfort and resistance often exhibited in new work environments because they knew exactly what to expect. They took ownership of it because they had been involved in decision-making throughout.

5.5   The value of design review

A summary of the participatory techniques used in design review on this project is listed below. Using them enabled the feedback and evaluation loops that are integral to the BPE approach.

5.5.1   Team walk-through

As part of strategic planning (Phase 1), the first step towards creating a shared vision was a structured team walk-through of the existing facility. The Hypertherm Management Team and consultants toured the facility as a group, discussing the tasks of each work-group, pointing out difficulties and advantages with the present space, and commenting on each other's presentations.

5.5.2   Goal setting

As part of the programming process (Phase 2) members of the five-person Hypertherm Management Team and the CEO met weekly to arrive at consensus on project objectives in a series of 2-hour facilitated work sessions. The results defined the organizational changes envisioned, identified the qualities of the physical environment they wanted to create, and set priorities for communicating these decisions to employees.

5.5.3   Design guidelines

The three priorities for the new organization were team-based work, cross-functional communication and a supportive and attractive work environment. All design decisions were documented as guidelines for Phase 3, design, and distributed to all stakeholders. Programming and design review merged at this stage, and design decisions progressed in response to and in tandem with intensive feedback and evaluation work with staff.

5.5.4   Pre- and post-occupancy user survey

The client attached major importance to involving employees at all levels in the management of change. To embark on design review, a questionnaire survey about the physical conditions of work in the existing building started the process of employee involvement. After occupancy, the same questionnaire was redistributed to all employees, and the results compared. The results were used for fine-tuning the space, offering advice to employees who had difficulty adjusting, and demonstrating the overall positive effects of the new building.

5.5.5   Pilot testing/environmental simulation

As the designers moved forward into design development, the second stage of this phase was to appoint one new, cross-functional team to test out some of the new workspace concepts. A similar process of testing a simulation of a proposed innovation is described in Chapter 15. As well as providing regular feedback on office layout and the new furniture, the presence of this group demonstrated and communicated the principles of the new space design to the rest of the organization.

5.5.6   Communication

The need for communication was identified and respected throughout design review. The CEO and members of the management team spent generous amounts of time sharing design and planning decisions with team leaders and ensuring that these were in turn communicated to team members. Once the new cross-functional teams were formed, representatives of each team – called 'space planning coordinators' – were selected to work with the designers. They met regularly with the designers to learn about and provide feedback on design decisions, and they also collected information from their teams for use by the design team. Meetings were held with all coordinators, not separately with the coordinator of each team. This helped teams negotiate among themselves for scarce resources, and kept everyone aware of decisions that were being made, such as team locations in the new building.

5.5.7   Design charrette

The design review participation process culminated in a two and a half day charrette, organized by the designers at the Hypertherm site. 'Charrette' is the term architects use for round-the-clock work sessions before a final presentation. Drawings of the new buildings showing team location and desk layouts were produced as 'final' versions, and each team's space planning coordinator was asked to sign off on the drawing as a contract agreement indicating that they accepted it.

5.6   Conclusions

Involving employees in design review was necessary in order for them to learn about the new, team-based organization, as well as to prepare them for the new workspace. Identifying space planning coordinators helped structure this involvement and make it real. The client required employee buy-in and ownership of change in order for managers to move ahead with their vision. The space planning process became the mechanism for enabling this to occur successfully.

As this example shows, design review is a key feedback loop in the design phase of BPE, and one in which it is easy to fail. In all too many design projects, designers react passively to programme requirements, and generate ideas that are later shot down by the client because a key piece of information was not included – either because users did not mention it during programming, or because designers accorded it a weaker priority. The balance that the designers have to establish between communicating their own ideas for solving the client's problem, and being sensitive to the priorities and preoccupations of the client, is a difficult and delicate manoeuvre. On the other hand, when it is successful, the design review loop ensures good design, satisfies clients and users, solves technical and financial problems early on, and results in a cost-saving process, as well as a good building.

In the case of Hypertherm, the benefits of the feedback and evaluation process accrued to designers, client and the construction team as well as to employees. Users were pleased with the new space, and were well prepared to occupy it and make it work. This example also indicates the value of design process facilitation, a role that can be performed by the strategic planner or programmer, the architect, or the project or facilities manager. However, the value of involving an unbiased professional, trained to manage group processes, cannot be underestimated when implementing feedback loops such as design review.

In summing up design review, it is clear that while feedback, evaluation and the acquisition of information to improve decision-making are all critically important to the creation of good quality buildings in today's world, it is important not to forget the immeasurably valuable and very personal contribution of those carrying out the creative act of design.

References

Preiser, W.F.E. and Schramm, U. (1997). Building performance evaluation. In *Time-Saver Standards: Architectural Design Data* (D. Watson et al., eds), pp. 233–238. McGraw-Hill.

Scheer, B.C. and Preiser, W.F.E. (eds) (1994). *Design Review: Challenging Urban Esthetic Control.* Chapman & Hall.

Vischer, J.C. (1995). Strategic Workspace Planning. *Sloan Management Review*, Vol. 37, **1**, Fall.

Vischer, J.C. (1999). Case Study: Can This Open Space Work? *Harvard Business Review*, May–June.

Vischer, J.C. (2003). *Work Environment and Well-Being: Beyond Working Space* (paper presented at Art and Synergy Through Design conference, Sydney, Australia: February).

Zeisel, J. (forthcoming). *Inquiry by Design* (2nd edn).

6

Phase 4: Construction – commissioning

Michael J. Holtz

Editorial comment

This chapter deals with Phase 4 of the framework for building performance evaluation, i.e. construction and the commissioning process, which constitutes the review loop of the construction phase. Architecture at its essence is about performance and the expression of human aspirations – the art and science of building. Architects must design buildings that meet the client's needs and expectations (basic performance requirements, such as spatial-functional needs and environmental control) within the local and regional context of the site, on time and within budget, in a visually pleasing manner. Architects want happy occupants and happy owners!

Architects capture and express the performance of their proposed design in the design and construction documents they produce. The expected and anticipated performance of the building – from a building systems point of view – is embedded in the architectural and engineering drawings and construction specifications prepared during the design process. The construction documents reflect the analyses performed and design decisions made during the design process, and the design team's collective design experience.

General contractors and their specialty subcontractors are contractually responsible to the owner to faithfully and accurately execute the design and construction documents. Their job is to build what is drawn and specified.

Given the current manner in which the design and construction processes are implemented, who is responsible for ensuring that the designed performance is achieved in the occupied, operating facility? It is in large part due to the unsatisfactory answer to this question that building commissioning services are becoming widely used in the building delivery process.

6.1 Introduction

From ancient times through the Renaissance, architects were 'master builders' responsible for the totality of the building delivery process. Architects coordinated overall planning, at both the community and building site scales, and were responsible for the architectural and engineering design of individual buildings, including interiors, landscaping, and furnishings. Clients entrusted architects with complete authority to carry out their planning and architectural design solutions by assigning them full responsibility for construction, including hiring of all craftsmen, approving all material choices, and supervising and approving construction quality. In essence, architects as 'master builders' were responsible for the ultimate performance of the completed and occupied design solution. As such, their clients held them personally responsible for the success or failure of the project, and many architects suffered the consequences of their failures.

Today, a number of intervening relationships come between the client (needs and expectations) and the architect (execution of the contract). Agents commonly intervening in this process include the owner or the owner's technical representative, the construction manager or design-build developer/general contractor, and specialists at every phase of the planning, design and construction process. Financial consultants, code consultants, engineers of every type, transportation specialists, building automation consultants and contractors, and others are now part of most design and construction processes. As a result of this myriad of players, roles and responsibilities can become confused or obscured, even with lengthy contracts between the various parties.

Out of this complex and confusing situation, planning is completed and buildings are constructed; however, the outcome of this process in terms of building performance has generally not been satisfactory. Buildings, and the systems which make up buildings, such as structure, fenestration, mechanical, electrical, control, communication, circulation, etc., often fail to perform in a manner ensuring health, safety, productivity, comfort, and well-being for occupants. Sick building syndrome, comfort complaints, dysfunctional or wasted space, expensive repairs and corrective changes are symptomatic of buildings not performing as intended by their designers, or as expected by owners and occupants.

6.2 Commissioning defined

The Building Commissioning Association defines the purpose of commissioning as follows: 'To provide documented confirmation that building systems function in compliance with criteria set forth in the project documents to satisfy the owner's operational needs' Building Commissioning Association, 1999). The essential elements of commissioning parallel phases of the BPE framework, and include:

- Reference – this is usually the owner's requirements, as documented in Design Intent and Basis of Design reports.
- Evaluation criteria – specific performance metrics must be established in terms of function and performance, based on the Design Intent and Basis of Design documents for those building systems to be commissioned.
- Commissioning activities – observation, verification/testing, and documentation activities are undertaken and implemented during the design, construction, and occupancy process.

Four different types of commissioning can be undertaken.

- Commissioning – process applied to new construction and major building renovation projects.

- Retro-commissioning – performed on facilities that have been in service and were never previously commissioned.
- Re-commissioning – facilities previously commissioned and in need of a 'tune-up'.
- Continuous commissioning – ongoing programme of structured commissioning throughout the lifetime of a building.

Fundamentally, building commissioning is concerned with the following question: Does the building, or specifically the commissioned systems, work as intended (i.e. as designed)? As such, commissioning is an important feedback loop in the BPE process (Preiser and Schramm, 1997). Occurring later on in the building delivery process than planning, programming and design review, as part of readying a building for occupancy (Phase 4 of building performance evaluation), commissioning nevertheless provides a continuous feedback loop to each of these phases.

6.3   Commissioning versus construction administration

Owners, and many designers, ask: 'Aren't I already getting/providing these services? After all, I have hired a design team and a construction team to deliver a completed, functioning facility. Why should I now have to pay for additional commissioning services?' The answer is that in the complex world of today's building industry, extra steps are necessary to ensure not only that technical and legal requirements have been respected, but also that the quality of the final product lives up to expectations. The conventional American Institute of Architects (AIA) design and construction administration contracts between owner and architect, and between owner and general contractor, do not address the purpose and goals of commissioning.

The definition and scope of construction administration services contained in the AIA Standard Contract B141 and Standard Contract B141 calls for activities to answer the following question: 'Is the building being constructed according to the contract documents?' (American Institute of Architects, 1997). Construction administration is primarily concerned with representing the owner relative to the integrity of the design: what was designed got built! Commissioning, on the other hand, is primarily concerned with ensuring that the intended (designed) performance has been achieved in the constructed and occupied building: the building works as intended! Construction administration and commissioning are different sides of the same coin. They address quantitative and qualitative criteria for quality control, respectively.

6.4   The commissioning process

The commissioning process is typically organized into four distinct phases, as shown in Figure 6.1.

Within each phase, a set of interrelated activities are performed that are designed to achieve the overall commissioning goals of the project. Typical commissioning goals, and a description of the activities undertaken in each phase of the commissioning process, are presented in this section. These are:

- Commissioning goals
  - Ensure occupant safety and satisfaction
    - All commissioned building systems performing reliably and safely

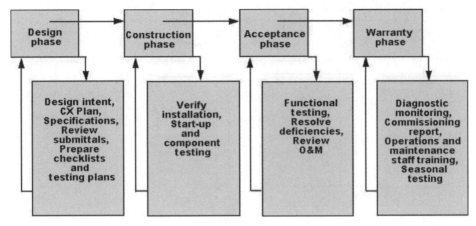

Figure 6.1 Four phases of the commissioning process.

- ○ Acceptable environmental comfort and indoor air quality consistently and uniformly provided
- – Achieve anticipated cost savings
- ○ Achieve project energy savings
- ○ Reduce equipment maintenance and equipment replacement costs
- – Achieve a smoother construction process
- ○ Avoid (minimize) construction-related problems
- ○ Achieve faster final acceptance and occupancy
- ○ Reduce warranty period callbacks
- ● Commissioning Objectives
- – Ensure the facility meets performance requirements
- – Provide a safe and healthy environment
- – Provide optimum energy performance
- – Provide a facility that can be efficiently operated and maintained
- – Provide complete orientation and training for the facility operations staff
- – Provide improved documentation of building systems
- ● Define future operations and maintenance (O&M) and continuous commissioning requirements.

6.4.1   Design phase commissioning

An effective building commissioning process begins during the design phase of the project, possibly as early as programming/briefing. The intent is to review the proposed design from an operations, maintenance and performance perspective, and to identify and resolve issues 'on paper' before they become actual physical (construction) or operation problems 'in the field.' Early intervention is always more effective and less costly than post-occupancy problem solving.

A key aspect of design phase commissioning is defining the owner's Design Intent and the design team's Basis of Design. These two documents establish the point of reference from which all commissioning activities emanate. The subsequent design reviews,

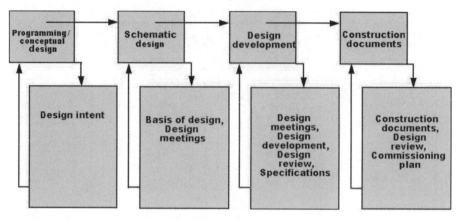

Figure 6.2

functional performance tests, and short-term diagnostic testing are all undertaken to determine if the design intent has been achieved in the constructed and operating system (see Figure 6.2).

The Design Intent report documents the owner's design requirements, establishes a baseline of performance expectations, and provides an explanation of important ideas, concepts, and specific performance criteria. The information included in the design intent report is integral to the architectural programming effort (see Chapter 4), and thus the owner's involvement is essential to define fully their performance requirements. The focus of these reports is typically on indoor environmental conditions which affect occupants' comfort, such as thermal comfort, luminous environment, acoustics, and indoor air quality, although broader building issues, such as building energy efficiency, material recycling, and reduced environmental impact, can also be addressed.

The Basis of Design report documents the assumptions behind the design team's decisions. It defines the design team's collective response to the owner's requirements/design intent. An example of a Basis of Design statement is as follows:

- Design intent: 'Ensure satisfactory indoor air quality'
  - Basis of design: 'The mechanical systems will follow the recommendations of ASHRAE Standard 62–1999 (ASHRAE, 1999). Zone-level carbon dioxide measurements will be used to establish occupancy levels. Ventilation air volume will be adjusted for current occupancy and verified by airflow measuring stations in the outside air intake.'

The Basis of Design report establishes the 'evaluation criteria' that will be used during the commissioning of the specific building systems – similar to the performance criteria described in Chapters 1 and 2 – as well as the types of commissioning activities and tests that will be required to demonstrate compliance with the owner's design intent.

Other key design phase commissioning activities include incorporating commissioning specifications in the contract documents, conducting commissioning review of design/construction documents, and preparing a preliminary commissioning plan, including prefunctional checklists and functional test plans.

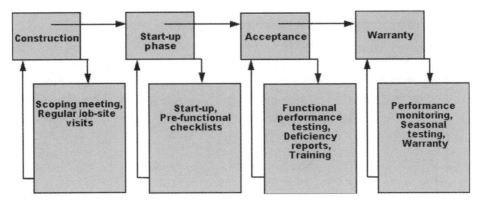

Figure 6.3

6.4.2 Construction phase commissioning

Construction phase commissioning initiates interaction between the commissioning agent and the construction team, which usually comprises the general contractor and the specialty subcontractors responsible for executing the project. Construction phase commissioning activities are divided into two segments – construction and start-up – as diagrammatically shown in Figure 6.3.

The primary commissioning activities undertaken during this phase are as follows:

- Revise the commissioning plan as required.
- Hold the commissioning scoping meeting, involving the owner, design team, construction team, and commissioning agent.
- Prepare final, pre-functional inspection checklists for the building systems to be commissioned.
- Verify proper installation of the equipment/systems to be commissioned.
- Observe start-up and component testing.
- Identify and resolve problems or deficiencies found during the installation and start-up process.

The primary emphasis during this phase is on proper installation and start-up of the building systems to be commissioned. Design intent and operational performance cannot be achieved if the building components and systems are not installed properly. Thus, pre-functional checklists and observation during start-up are completed to demonstrate that the systems are now ready for functional performance testing.

6.4.3 Acceptance phase commissioning

Functional performance testing is the primary activity undertaken during the acceptance phase of the commissioning process, and is often referred to as the 'commissioning' phase in the general contractor's project schedule (see Figure 6.4). Functional performance tests verify the intended operation of systems components and associated controls under various conditions and modes of operation, although the building is still at this point unoccupied.

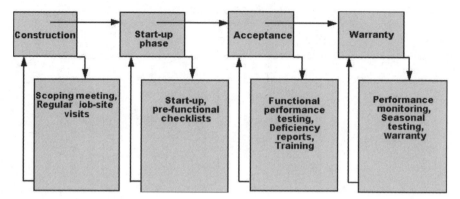

Figure 6.4

Functional performance test plans are prepared for each specific building system component and the complete building system. They require extensive coordination with multiple subcontractors, as well as testing, adjusting, and balancing (TAB) to ensure that contractors responsible for the performance of each building system participate in the diagnostic testing. Results from the functional performance tests are incorporated into 'punch lists' and commissioning 'deficiency reports', which identify problems that must be corrected before final completion and occupancy.

The building's operation and maintenance (O&M) personnel should be involved in and observe all functional performance tests. This is often when O&M staff learn how the building systems are supposed to operate relative to the original design intent. Additionally, they learn the preventive maintenance procedures needed to ensure effective operation of the building systems over time. This hands-on involvement of the O&M staff during functional performance testing supplements the normal operation and maintenance training provided by the general contractor and the specialty subcontractors, as required in the construction contract.

Upon verification that all 'punch list' and commissioning deficiencies have been resolved to the satisfaction of the design team and the commissioning agent, the building is officially turned over to the owner, and the general contractor and subcontractors leave the site.

6.4.4   Warranty phase commissioning

To some, commissioning ends at the conclusion to the acceptance phase. However, at this stage the building has not been tested while occupied. Functional performance testing has only verified operation, not performance, of the targeted building systems, thus the experience of people working in the building can only be estimated.

Warranty phase commissioning involves short-term (two to three weeks) diagnostic monitoring of the building systems under normal occupancy and operating conditions. This type of diagnostic testing investigates the dynamic interactions between building system components while the building is in use. Battery-powered data-loggers, or the building automation system trend-logging capability, are used to record undisturbed

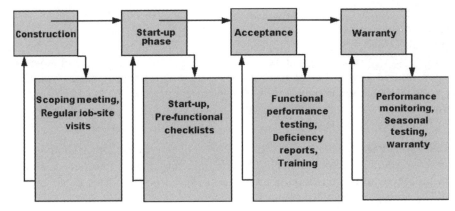

Figure 6.5

system operation. Analysis of these data by the commissioning agent will identify faults difficult to detect during functional performance testing.

Short-term diagnostic monitoring is typically performed several times during the warranty phase to address seasonal operation variations – summer, winter, and the spring seasons. The operation and performance problems identified during the warranty period are still the responsibility of the general contractor and specialty subcontractors. The O&M staff also benefit from the short-term monitoring/diagnostic testing because the various building systems' performance can be 'benchmarked', and thus future performance of the systems can be compared to the original, correct conditions of operation. In the best of situations, the warranty phase gives way to a continuous commissioning phase where the building systems are constantly monitored and evaluated, so that their operation and maintenance ensures energy savings and operational reliability.

A final commissioning report is prepared by the commissioning agent, which summarizes all findings, conclusions, results, and recommendations from the commissioning process. It becomes part of the owner's building and construction documentation and is a key element of the facility's O&M staff documentation, and therefore a tool for facilities management (as described in Chapter 8). If properly written and used, the commissioning report provides a technical basis for operating and maintaining the building at peak performance for its entire lifetime. Commissioning is therefore a loop, applying not only to the occupancy phase, but also to the facilities management phase of building performance evaluation. The final commissioning report can take up several volumes, including a building user manual to inform and assist building occupants and O&M staff in maintaining a safe and comfortable environment.

6.5   Commissioning tools

Commissioning is a knowledge-intensive process. The commissioning agent must be knowledgeable about many aspects of the building design, construction, operation, and maintenance process. The commissioning agent must also have excellent communication and interpersonal relationship skills, because the job involves constant interaction with the design team, owner's representative, general contractor, specialty subcontractors, equipment representatives, code officials, and so on. Commissioning agents must be able to adapt to the unique needs of each project. Most commissioning agents operate on the philosophy of 'solve problems – don't place blame!' All parties want a successful project and a happy owner, so typically all parties can and should work together to identify and resolve the problems that arise throughout the building delivery process.

The commissioning agent relies on a number of tools. A brief summary of the key commissioning tools is presented below:

- Commissioning database – a software database tool to record and track all information generated during the commissioning process, including contact information, commissioned equipment information, pre-functional checklists, short-term monitored data, deficiency reports, and final commissioning report.
- Commissioning plan – a document that defines the scope of the commissioning effort, the roles and responsibilities of all parties, pre-functional checklists and functional performance test plan, and short-term diagnostic monitoring requirements.
- Pre-functional checklists – unique checklists for each piece of equipment to be commissioned, describing the installation and start-up procedures that must be performed and verified by the installing contractor prior to functional performance testing.
- Functional performance test plans – unique test plans for each piece of equipment or a complete building system to be commissioned, describing the test protocol to be used to assess the functional performance of the equipment or building system.
- Test equipment – test equipment, such as data-loggers, sensors, hand-held meters, digital cameras, used during the commissioning process to test or monitor the installation, operation and performance of the building systems being commissioned.
- Deficiency report – an ongoing record of the faults (deficiencies) found during the entirety of the commissioning process, the corrective actions required, the parties responsible for the corrective action, and the final result/resolution of the deficiency.

6.6   Conclusions

Building commissioning is a formal process to ensure and document that building systems perform in accordance with design intent and owner expectations. Taking place further on in the BPE process, that is to say, at the moment of occupancy, commissioning provides an important feedback loop linking users and managers of buildings with their owners, designers and builders. It is distinct from, but complementary to, traditional construction administration services. It is a way of ensuring quality control and protecting the ultimate user or occupant from unsafe or unsanitary conditions, both at the moment of occupancy and over the lifetime of the building. As the complexity and integration of building systems increases, building commissioning will become an essential activity within the building delivery process.

References

The American Institute of Architects (1997). *AIA Standard Contract B 141, Standard Form of Agreement Between Owner and Architect*. Washington, DC.

American Society of Heating, Refrigeration, and Air-Conditioning Engineers (1999). *ASHRAE Standard 62-1999.*

Building Commissioning Association (1999). 'Building Commissioning Attributes', web document revised April 14.

Preiser, W.F.E. and Schramm, U. (1997). Building performance evaluation. In *Time-Saver Standards: Architectural Design Data* (D. Watson et al., eds), pp. 233–238. McGraw-Hill.

7

Phase 5: Occupancy – post-occupancy evaluation

Bill Bordass and Adrian Leaman

Editorial comment

Post-occupancy performance evaluation tries to answer four broad questions: 'How is this building working?', 'Is it intended?', 'How can it be improved?' and 'How can future buildings be improved?' Answers to these vital questions can add value both to the buildings assessed, and to client, industry and management procedures for future building construction and renovation. In spite of the significant accumulation of knowledge based on user feedback studies and post-occupancy evaluation (POE), those who procure, design and construct buildings are still often reluctant to engage in or carry out POEs on a systematic basis.

This chapter explores steps that may be necessary if building performance assessment and feedback is to become a routine part of building delivery and the building's life cycle. Techniques need to be streamlined, making them more robust and relevant in application, bringing them closer to the real world, and providing easier access to methods, results, conclusions and support systems. If POE procedures are made quicker and cheaper, then they are more likely to be incorporated into procurement, briefing, management and professional development processes in simple, direct and timely ways.

These recommendations are illustrated by projects in which the authors have been involved – mainly from the UK but with international relevance. They include the evolution of occupant and energy surveys; the 'Probe' series of twenty published post-occupancy surveys; and new techniques for building performance assessment and feedback.

7.1   Introduction

This chapter is about looking at how buildings work in practice, in order to improve the performance of both the buildings examined, and the levels of knowledge and understanding of those who commission, design, build and manage them. In the past, designers and builders often lived close to their clients and occupiers, so inevitably received direct feedback on the performance of their completed buildings. In the twentieth century, the construction industry has evolved from being a crafts-based industry to more of a science, but the science is not yet all there. Although containing familiar elements, few buildings are identical; so in scientific terms each design decision is an hypothesis awaiting its experimental test. But where are the designer-experimenters? In the past, when technology and user requirements changed slowly, one could perhaps rely on evolutionary feedback and the occasional catastrophe. More recently, one may look to academic study and the test of time. But today and in the future, when things are changing so fast, there is really no alternative to 'learning on the job'. And how better than for the producers to be involved in post-occupancy evaluations of the buildings they have produced?

7.2   The origins of POE

7.2.1   POE in the UK

While building evaluation activities are not new, POE as we know it probably emerged from the trend to science-based building in the 1950s and 1960s. In its 'plan of work for design team operation', the Royal Institute of British Architects (RIBA, 1963) broke down the sequence of briefing/programming, design, specification, tendering, construction, completion and use into clearly-defined stages. This included a final Stage M (feedback), when the architects would examine the success of what they had done.

This intention to undertake feedback in the UK began to be given academic rigour in the late 1960s when twenty architectural and engineering practices, the Royal Institute of British Architects (RIBA), the *Architects' Journal* and the Ministry of Public Building and Works sponsored the Building Performance Research Unit (BPRU) at the University of Strathclyde to undertake feedback, to bring together research, teaching and design on building performance, and to publish the results. BPRU only lasted four years in this form. Its work was largely on schools. The results were published in the *Architects' Journal*, and in *Building Performance* (Markus et al., 1972), in which practical findings that are still true today were enumerated. The book included a plea for architects to be more involved in feedback.

Ironically, the year this book was published, the RIBA removed Stage M from its publication, reportedly because clients were not prepared to pay for feedback as an additional service, and the RIBA did not wish to create the impression that feedback would be undertaken as a matter of course. Today the wheel seems to have turned full circle, with the RIBA (1999) saying that 'the biggest improvement to be made (in customer focus) is in systemizing feedback and in instituting post-occupancy evaluation'. In 2003, forty years after Stage M first appeared, the RIBA Practice Committee decided to reintroduce it into its published documents. Although it is still rare for architects to be involved in routine feedback activities, interest is growing. A statement in Markus et al.'s (1972) book may indicate why evaluation feedback had not become mainstream: 'BPRU was more interested in research than in developing devices, however practical, without a sound theoretical framework.' This orientation may have distanced researchers from the designers, clients, operators and users they had intended to serve (Cooper, 2001).

7.2.2 POE in North America

In North America, the scientific approach to POE also seems to have started in the 1960s, though the term did not come to be applied until the 1970s, in military facilities. Van der Ryn and Silverstein (1961) write about buildings as hypotheses requiring experimental 'verification'. Their first set of verifications (or, strictly, refutations, as the occupants had lots to criticize in their award-winning student residences!) appeared in 1967. Soon afterwards, the Environmental Design Research Association (EDRA) was set up, and has helped to sustain the practice of POEs in North America, principally in the academic mould and concentrating on psychological aspects of user satisfaction.

As in Britain, it has proved less easy to connect POEs directly with design and construction activities. In 1987, a US federal committee found that although a few federal agencies did POEs routinely, there was a widespread lack of institutional support. The committee recommended making POEs more rigorous and systematic; laying the groundwork for a database on building use and performance; and establishing a clearinghouse to assemble, maintain and disseminate POE information. A new initiative is described in Chapter 17 of this volume, bringing together several public and private agencies for collecting, classifying and applying findings from POEs.

7.3 Recent developments

7.3.1 North America

The Federal Facilities Council has recently reviewed the situation in US federal agencies in a useful introduction to the state of current practice (FFC, 2001). The book suggests that management trends to focus and downsize organizations, to outsource services, and to concentrate on achieving strategic business goals have kindled a new interest in POE. However, having downsized their buildings' expertise and outsourced their essential feedback loops, the same organizations may lack the insights, skills and confidence to take effective action. The FFC concluded that POE was now essential to replace lost, in-house, often tacit and informal knowledge about buildings. The techniques required go beyond user satisfaction surveys to cover all activities that affect how a building performs. Fortunately, many such techniques are already available; several are presented in this book in the context of building performance evaluation.

7.3.2 United Kingdom

In 1994, a change in building research funding by the UK Government allowed an unusual team to put forward the idea of undertaking and publishing POEs on new buildings of technical interest, typically three years after completion. The project was led by the editor of *Building Services*, the house journal of the Chartered Institution of Building Services Engineers (CIBSE – the British equivalent of ASHRAE), with a multi-skilled team of individuals who bridged the gap between research and practice. To maintain their independence, the team made direct contact with owners and occupiers of the targeted buildings. The designers provided supporting information, and they and the occupiers were invited to comment on the findings before publication. Between 1995 and 2002, the Post-occupancy review of buildings and their engineering (Probe) team published twenty POEs and a variety of review papers, including a special issue of Building Research and Information (2001).

Probe demonstrated that one could put feedback information on named buildings into the public domain without falling victim to legal action, as some had feared. Its main quantitative tools addressed 'hard' issues (the TM22 energy survey method [CIBSE, 1999]) and 'soft' issues (the Building Use Studies' occupant questionnaire) – both of which provided reference benchmarks. With the associated interviews, walk-through surveys and reviews of technical background information, the results of these two surveys overlap into other areas, e.g. briefing/programming, procurement, build quality and business and facilities management, much like the processes described by Preiser and Schramm in their original presentation of Building Performance Evaluation (Preiser and Schramm, 1997).

Probe aimed to generate results directed towards designers and their clients, rather than specifically feeding back to the building in operation and the personnel concerned; however, comprehensive reports were given to building managers. Many of them were already operating monitoring and feedback systems, and used the results to make further improvements. In other cases, some cultural change within the organization was necessary to ensure that feedback was applied to routine design, construction, procurement and management practices.

7.4   Making feedback and POE routine

7.4.1   Conclusions from Probe

- An important conclusion from Probe was that feedback from building users could help to add value without increasing cost, by linking means (the constructed facility) more closely to the client's ends. Additional recommendations include the following:
  - Clients should define their objectives more clearly, and undertake monitoring and reality-checks throughout the design and construction process.
  - Designers should get to know more about how buildings really work from a technical perspective and make them better, more robust, more usable and more manageable.
  - The supply side (construction industry) should establish 'no surprises' standards and provide support after handover, for example, with provision for 'sea trials' periods in standard contracts, with better proving of system performance and provision of after-care to clients and occupants (see Chapter 6 on the commissioning processes).
  - Facilities managers should monitor more carefully, be more responsive to user needs and comments, and represent client requirements more strongly (see Chapter 8 on facilities management).
  - Professional institutions should encourage feedback procedures as part of normal practice.
  - Government should encourage feedback among other measures that lead to all-round improvement.

7.4.2   Involving construction clients

The UK government has for a long time been unhappy with the poor efficiency and value for money offered by the construction industry. In the early 1990s, the Conservative government commissioned a report (Latham, 1994), which put particular stress on team building. Not so long afterwards, the new Labour government commissioned a new report (Egan, 1998), which stressed efficient supply chain management. The Confederation of Construction Clients (CCC) was launched in 2000 to set up mechanisms for ensuring better feedback from completed buildings. Up to this point, initial research had made it clear that it would be difficult to get clients to adopt and pay directly for a feedback system. As well as a purported lack of interest from senior management, many clients knew little about the range of techniques available, how they should be used, and what value they might add. They were concerned that any added value might accrue not to them, but to the designers' future clients; or even that their efforts would be wasted as a result of poor feedback loops and knowledge management systems within individual organizations and even throughout the whole industry.

Clients and designers were also unsure of the immediate payback of routine evaluation phases and loops. This impression was reinforced by the name post-occupancy evaluation: an activity that some facilities managers did not consider to be a dynamic contribution to continuous improvement!

In spite of some interest in post-completion checks to ensure that buildings complied with design requirements, clients tend to be more oriented to project sign-off checklists rather than any real engagement with the in-use performance of the buildings in the sense of commissioning (Phase 4 of BPE) and evaluation (Phase 6). In addition, client requirements are not always explicitly stated in the programming phase, such that they can be checked off systematically upon occupancy (one of the casualties of today's hurried world has been a lack of time for a thorough process of briefing/programming – see Chapter 4). Even where requirements were clear initially, they were not necessarily carried through to the end of the project, and change orders and their consequences were not incorporated into formal revisions of the brief or the design.

7.5   Moving forward

7.5.1   Design practice

The Probe findings and apparent loss of construction clients' enthusiasm for evaluation led us to look to design and building teams for further support in making feedback routine. Designers proved to be a more homogeneous interest group, involved in a range of different procurement systems, and more used both to collaborating to provide professional competencies and to competing for excellence. Historically, as the experience with RIBA Stage M described above shows, designers have stayed away from routine POE and building feedback. When a project is completed, funds are often short and the next project imminent. Some are also concerned about their responsibilities if things are not working, as well as the effects on their insurance.

Nevertheless, in recent years interest in building performance in the UK has been growing. Specifically, Government clients have been encouraged to specify by output rather than input requirements, in particular for private finance initiative (PFI) projects. Various organizations (including the recently-formed CABE – the Commission for Architecture and the Built Environment) want to improve the design quality of public buildings and spaces. Moreover, the European Union is now requiring (OJEC, 2003) energy labelling of buildings, which promises to reduce the gaps between predicted and in-use building performance.

Leading design firms are therefore coming to realize that to develop a better understanding of how their buildings actually perform in use and over time is essential to their survival. We are therefore working with a user group of twelve firms to look at ways of making feedback a routine part of project delivery. As the FFC (2001) study recommended against a single preferred method of POE because contexts, needs and resources could vary, our approach has been to develop a portfolio of feedback resources. This currently includes techniques for undertaking feedback and POEs, and will evolve to contain results, contacts, guidance notes and more.

7.5.2   Techniques

In order to meet previously-stated criteria of speed, accuracy, effectiveness and cost-effectiveness, experience gained through the development of the CIBSE TM22 energy assessment method and the building use studies occupant survey (see Appendix) indicates that feedback techniques must meet the following criteria.

Feedback (and POE) resources must:

- appeal to a wide spectrum of clients;
- be applicable in a range of building types;
- be comprehensive in the details that they cover;
- be as simple as possible, but not simplistic;
- be practical, with a real-world (not theoretical) emphasis;
- be rapid to administer on site, with speedy turn-round of results;
- be acceptable to building managers so that normal use of a building is not unduly hindered;
- be capable of dealing with subtle changes from one building and commissioning client to the next;
- provide unambiguous factual data which are well presented and easy to interpret;
- be relatively cheap;
- be based on a robust core methodology which meets stringent criteria from different standpoints, e.g. academic (for hypothesis testing), design practitioner (diagnostics and feedback assessment) and facility management (e.g. change management);
- have continuity, and not fall by the wayside after the development phase is over;
- have, where possible, capability for application internationally.

Techniques that are beginning to meet most of these criteria tend to use standard software products, and extend them with new features that are purpose-written (e.g. statistical graphics for benchmarking and certification); and to use the Internet appropriately (e.g. for dissemination of results, rather than data gathering). Several have simple but powerful analysis databases.

If feedback is to become a routine part of project delivery, as Preiser and Schramm point out in Chapter 2, techniques are most appropriately embedded in a process that can be managed through the life cycle of a building. The portfolio of feedback resources classifies the applicability of techniques according to ten key stages, considered in five blocks of two each (see www.usablebuildings.co.uk/fp/index.html).

More techniques could eventually be added. However, the user group set up to test the techniques has started with a small number of (where possible) well-proven techniques, in order to build up a record of experience, and benchmarks that will help to reinforce their acceptability. Many of these techniques could function as feedback loops in the BPE framework.

7.6   Conclusions and next steps

The conditions are now right for making POE and other feedback loops in the BPE framework routine. Clients are becoming more interested in building performance. A growing number of techniques are available to help them, some with long track records. IT and the Internet are making these techniques faster, more powerful, easier to use, and more economical; they are backed up in some instances with databases of statistics and benchmarks.

Obstacles still remaining include the following:

● Litigation: to date, the supply side of the industry has not routinely measured the performance of its products, so routine performance evaluation will inevitably reveal shortcomings. Unfortunately, this could feed litigation. Getting feedback systems going must be seen as a learning experience and a route to rapid improvement, not as an opportunity to place blame on the brave pioneers.
● Financing: more and more organizations are interested in the benefits of feedback, but few of them want to pay for it. Even though feedback activities are highly affordable in principle – because they make other activities more efficient and less wasteful – they still have to be programmed and budgeted for.
● Bureaucracy: performance assessment systems are often developed by and for officials. It is vital that any feedback systems imposed from above are derived from things which have been found practical and useful by those actually working on projects, as clients, designers, users, builders and managers.
● Knowledge management: the data produced from POEs need to be managed in order to lead to effective learning. Knowledge management systems are undeveloped in most firms of designers, builders, and construction and maintenance departments. This is one reason why recent efforts have concentrated on project teams and their immediate clients, who are able to put their experience and new understanding into action immediately.

In the UK, the relatively new Usable Buildings Trust (established in 2002 to promote better buildings through the use of feedback), was greeted enthusiastically by practitioners and industry, and may point the way forward to the coordination, development and promotion of systematic building feedback to give us better, nicer, more productive, more cost-effective and more sustainable buildings.

References and further reading

Baird, G., Gray, J., Kernohan, D. and McIndoe, G. (eds) (1996). *Building Evaluation Techniques*. McGraw-Hill.

Building Research and Information (2001). Special Issue on post-occupancy evaluation, *Building Research and Information* **29**(2), 158–163, March–April.

CIBSE (1999). *Technical Memorandum TM22:1999, Energy assessment and reporting methodology: office assessment method.* Chartered Institution of Building Services Engineers, London.

Clark, P. (2003). Handover heart, *Building*, **57**, 13 June.

The Colonial Office (1909). Correspondence (April 1906 to January 1909) respecting the design of bungalows provided for government officials in East Africa. Colonial Office, African (West) No. 848, London, March.

Cooper, I. (2001). Post-occupancy evaluation – where are you? *Building Research and Information*, **29**(2), 158–163, March–April.

Egan, J. (1998). Rethinking construction: the report of the Construction Task Force to the Deputy Prime Minister, John Prescott, on the scope for improving the quality and efficiency of UK construction. DETR, London.

Federal Facilities Council (2001). Technical Report 145, Learning from our buildings: a state-of-the-practice summary of post-occupancy evaluation. National Academy Press, Washington.

Jaunzens, D., Grigg, P., Cohen, R., Watson, M. and Picton, E. (2003). Building performance feedback: getting started. *Building Research Establishment Digest* 478, BRE Bookshop, London, June.

Latham, M. (1994). Constructing the team: final report of the government/industry review of procurement and contractual arrangements in the UK Construction Industry. DETR, London.

Levermore, G.J. (1994). Occupants' assessments of indoor environments: questionnaire and rating score method. *Building Services Engineering Research and Technology* **15**(2), 113–118.

Markus, T., Whyman, P., Morgan, J., Whitton, D., Maver, T., Canter, D. and Fleming, J. (1972). *Building Performance.* Applied science publishers, London.

OJEC (2003). Directive 2002/91/EC of the European Parliament and of the Council of 16 December 2002 on the energy performance of buildings. *Official Journal of the European Communities I'65*, 1 January.

Preiser, W.F.E. and Schramm, U. (1997). Building performance evaluation. In *Time-Saver Standards: Architectural Design Data* (D. Watson, M.J. Crosbie, and J.H. Callender, eds), pp. 233–238. McGraw-Hill.

RIBA (1963). *Plan of Work for Design Team Operation.* RIBA, London.

RIBA (1999). *Architects and the Changing Construction Industry.* RIBA Practice, London.

Van der Ryn, S. (1961). *AIA Journal.* Full details not available.

Van der Ryn, S. and Silvestein, M. (1967). *Dorms at Berkeley: the ecology of student housing.* Educational Facilities Laboratories, New York.

8

Phase 6: Adaptive reuse/recycling – market needs assessment

Danny S.S. Then

Editorial comment

Business performance is contingent upon effective use and management of all resources to enhance competitive advantage. However, while the resource value of finance, human resources, and technology is widely recognized, that of the supporting building infrastructure (or real estate) that houses these resources is not obvious to many corporate managers, who see building-related expenses as a drain on profit.

Operational buildings are simultaneously a physical asset, a functional facility, and a business resource. It follows, therefore, that the measurement of building performance during occupancy will be multidimensional. The published literature suggests a wide range of views, which tend to polarize either towards the measurement of physical (technical) performance or of financial (cost) performance. Recently, a business resource view has emerged, which focuses on measuring performance in terms of the relationship between operational facilities and business outcomes. This view coincides with the facility management approach to optimizing building performance over the life of the building, and using feedback from a wide range of sources to ensure responsiveness to market needs.

The building performance evaluation (BPE) process encompasses design and the technical performance of buildings in reference to human and social (market) needs. This chapter discusses ways of assessing facility performance such that useful information can be applied to adapting buildings to change and making decisions regarding their future use.

8.1   The drive for measures of building performance

Buildings provide environments for a wide range of business activities. Some building facilities are a critical and integral part of the business (e.g. hotels, research laboratories, prisons, hospitals and schools), while others are more generic business resources (e.g. office space). A building represents a unique resource, in that it is a capital asset that requires a long-term funding decision, and, at the same time, an operational asset used by organizations as a factor of production. As a productive environment for business enterprises, facilities are often the largest asset group an organization may have, and the second largest expense after staff costs (Morgan, 2002). Under the right economic conditions, they provide capital appreciation as well as a recurrent income stream. Conversely, without a proper management and maintenance regime and good customer service, buildings can progressively deteriorate in value and have a shortened effective economic life span. Buildings are also ubiquitous assets, appearing on the balance sheets of both the private and public sectors.

Competitive pressures and economic conditions are driving companies' search for competitive advantage beyond a focus on costs and budgets alone. Businesses and governments both need to develop an informed view of what customers and end-users of services value, and the level of their performance expectations (Leibfried, 1992). These business drivers have a direct influence on how buildings and facilities should perform. As a result, the drive exists to explore the performance of all business resources – people, property and technology (Then, 1994).

8.2   Building performance and facility management

According to the BPE framework, the facility management phase is the longest, and the need to monitor its ongoing performance as an operating resource and as a business expense is therefore pressing (Preiser and Schramm, 1997). Corporate interest in monitoring the performance of buildings as assets can be attributed to their unique characteristics, as follows:

- the capital-intensive nature of building assets (usually worth many millions of dollars which could potentially be applied more profitably elsewhere);
- their durable nature (often lasting up to 20–50 years or more);
- their relative inflexibility in responding to changes in business directions and technology;
- the significant accompanying stream of recurrent expenditure burden associated with maintaining and operating them at a desired service standard;
- the potential liabilities due to deterioration and depreciation over time;
- their impact on productivity and business performance;
- their vulnerability to a wide range of legal requirements and risks.

The importance of property performance measurement in management practice is well recognized (Becker, 1990; Bon et al., 1994). However, the practical implementation of a performance measurement system such as that outlined in Chapter 2, that delivers the desired management outcomes efficiently and effectively, is more problematic (Tan et al., 2000). Measuring performance in facility management is obviously not an easy task, as there are different stakeholders with different interpretations of what is actually being delivered.

Some facility managers may argue that their task is protecting the owner's investments by managing facilities in the most cost-effective manner. In this case, performance may be measured through the increase in value of the facilities, or through the profit margin that they have been able to deliver to the owner. In the industrial or commercial sectors, their performance may be measured by level of production or sales, as compared to operation costs. Others may apply measures of staff productivity in order to gauge the effectiveness or success of facility management. In all cases, performance measurement requires accurate feedback on the efficiency and effectiveness of facility operations. Efficiency and cost have always been the primary focus in facility management; however, effective management of space and related assets for people and processes has assumed greater importance in recent years.

8.3 Business context of operational facilities performance

The rationale behind the drive to measure building performance is the realization that the costs of building construction represents only a fraction of the overall costs of ownership (see Chapter 11), and that building assets are a means to an end – they are the infrastructure support (operational facilities) required for the delivery of accommodation services. The acknowledgement that operational buildings are durable physical assets that require ongoing management has created a need for more robust decision-making tools that are capable of evaluating the respective influences of different operations, maintenance and service usage strategies throughout the extended service life of a facility. If maintenance investment is inadequate, or higher usage imposes greater stress or wear on the facility than anticipated, life cycle costs are likely to increase, adaptation and reuse possibilities are more limited, and the building's effective life shortened.

To be useful to management, building performance information must enable managers to act on it. Action can occur at two levels. At the portfolio management level, decisions on further action or investigations could be triggered by a limited number of key indicators that represent the 'vital signs' (Hornec, 1993) of the 'health' of corporate operational facilities. These key performance indicators (KPIs) need to be meaningful, few in number and have the capability to point to specific areas for action or further investigation and reporting.

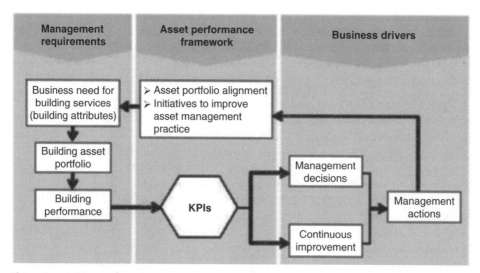

Figure 8.1 Building performance as catalyst for management action.

For facilities management, other measures are also required to perform diagnostic and continuous improvement roles (addressing specific areas of concern), or to throw light on causes of problems.

At a practical level, the choice of performance indicators is driven by management reporting requirements, whereas decisions on the selection of KPIs are shaped by the circumstances and context in which performance measurement is occurring. Management decisions and actions for continuous improvement and portfolio adjustment are influenced by business drivers, but rely on performance information to effectively identify target areas. Figure 8.1 illustrates the context and role of a conceptual performance framework in which building performance acts as the catalyst for management actions.

8.4   Review of current practice in measuring facilities performance

A wide range of methods and frameworks for performance measurement of building assets has been proposed (McDougall et al., 2002). These range from the detailed technical assessments of physical aspects of buildings (DeJonge and Gray, 1996; Damen and Quah, 1998) to surveys of user satisfaction with the occupied space and quality of the internal environment (Preiser, 2001; Szigeti and Davis, 2001; Vischer, 1996). Over the last decade or so, there is a wider acceptance that the workplace environment has an impact on productivity (Leaman and Bordass, 1999; Drake, 2002). The issue of user satisfaction in terms of fitness for purpose and quality of workplace environment has added an additional dimension to the concept of market needs analysis. In the wider context of the property and facilities industry, there are indications of a coming together of real estate provision and facilities services management as a result of restructuring of the supply side of the market (Then, 2002; O'Mara, 2002). This trend is also evident from the maturing of the integrated service delivery market, with service providers offering bundled services that may include construction, maintenance, cleaning, security and office services (Asson, 2002; Varcoe, 2002). From the perspective of corporate management there is now a closer scrutiny of occupancy costs (Apgar, 1993) and a willingness to consider alternatives to in-house provision for a range of facility support services (McBlaine and Moritz, 2002).

Published literature on performance criteria for facility management indicates a lack of consistency in the concept of asset performance as applied to operational buildings as a class of durable assets. At the same time, there is also general recognition that measurement of process performance should cover the needs and expectations of end users in addition to efficient facility provision. However, while the need for a balance between facility 'efficiency' and facility 'effectiveness' is well understood (Kaplan and Norton, 1996), the process of selecting the right set of performance criteria to achieve a balanced approach is not so well defined. One idea is to shift focus from measuring resource inputs to measuring outputs and outcomes (Price, 2000). The challenge appears to be the selection of a select few performance indicators from a myriad of possible indicators available at the building level, that can be aggregated upwards to the facility and portfolio levels for corporate management monitoring and alignment of building utilization.

Figure 8.2 illustrates a 'mental map' of the business context of building performance evaluation. The six questions provide a basis for considering both stakeholders' and corporate management's parameters, justifying a structured approach to developing a building

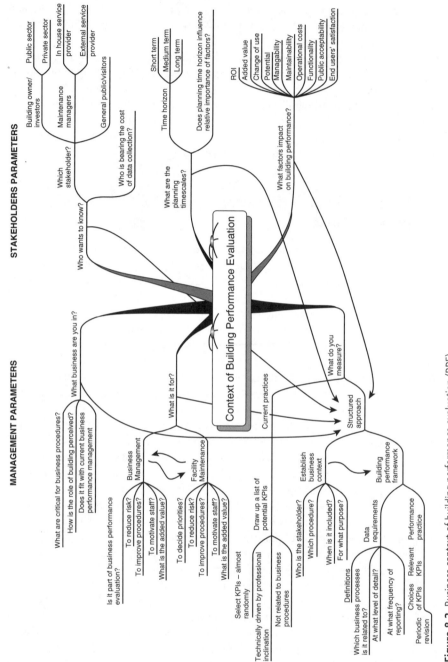

Figure 8.2 Business context of building performance evaluation (BPE).

performance framework. The stakeholders' perspective is critical in the selection of the appropriate few key performance indicators (KPIs) that provide the 'vital signs' (Hornec, 1993) necessary for management decisions and actions associated with operational facilities. The measurement of building performance is clearly complex in that, unless the purpose of measurement is clearly defined, there is a real risk of being overwhelmed by the potential number of available KPIs.

8.5   A new view for determining facilities performance

In viewing buildings as a business resource, asset and facility management practices should be targeted towards the provision of buildings and facilities that respond to business objectives. This provides a basis for measuring building performance that can be framed against the outcomes of asset and facility management practices, as illustrated in Figure 8.3. Building assets can then be monitored over time to track their performance in contributing to meeting business objectives, or benchmarked with others.

Table 8.1 illustrates key asset/facility management practices that focus on ensuring building assets fulfil business requirements and effectively contribute to achieving business objectives. A balanced, five-facet performance framework based on the above considerations is shown in Table 8.2.

Economic metrics provide performance information on the alignment of building assets with strategic business directions, in terms of their location, type, level of investment, utilization, quantity and quality (age, condition, technology, etc.) of assets. They are needed to ensure the provision of an economical and appropriate configuration of assets in alignment with business plans and market offerings.

Functional metrics provide performance information on the outcomes of management decisions that relate to the creation of assets that are 'fit for purpose' as well as providing the desired working environment congruent with the organizational culture and workplace standards. The objective of measurement is to ensure the continuous functional alignment of available space with business operational needs, including the match between demand and supply.

Physical metrics provide performance information on the effective and efficient management of operational aspects of ongoing asset management. The objective of measurement is the visibility of ownership and occupancy costs to ensure costs are affordable and commensurate with an appropriate level of risks and potential liabilities.

Figure 8.3 Linking building performance to business-driven asset/facility management.

Table 8.1 Business-driven asset and facility management practices (Then and Tan, 2004)

Asset and facility management practice	Management focus	Key elements
Strategic asset planning (business attributes)	Building assets as facilities in support of business delivery	• Integrated asset strategic plans (investment, management and disposal) driven by business plans • Policy guidelines on required real estate variables (location, design/form, flexibility, durability, etc.)
Facilities strategic brief (functional attributes)	Appropriate and productive working environment reflecting corporate culture	• Policy guidelines for workplace strategies and desired workplace environment • Policy guidelines on asset attributes and specific requirements • Policy guidelines on space standards
Facilities operation and management (service levels attributes)	Operation and maintenance of building assets (managing costs and affordability)	• Policy guidelines on standard of asset care and ongoing management monitoring • Policy guidelines on support services standards in terms of service levels and preferred procurement methods
Service standards specification (users' attributes)	Business operational needs and customer service standards	• Defined customer service standards (internal and external) • Specific standards related to business operations

Table 8.2 Facets of a balanced performance framework (Then and Tan, 2002)

Performance facets	Measurement focus
Economic metrics	Measures and indicators that reflect the economic alignment of building assets with business requirements in terms of costs, location, type, quantity and quality.
Functional metrics	Measures and indicators that reflect the 'fitness for purpose' of the building assets including considerations of an appropriate and productive working environment in terms of configuration, layout and amenities.
Physical metrics	Measures and indicators that reflect the physical integrity, condition and appearance of the building assets in terms of operating and maintenance costs.
Service metrics	Measures and indicators that reflect the satisfaction of users with the building assets in service as an operating facility.
Environmental metrics	Measures and indicators that reflect the wider role of building assets and their impact on the built environment at the levels of the natural ecology, the community as well as the specific operational facility. This facet may be governed by prescribed sustainability targets at project/state/national or international levels.

Service metrics provide information on the quality perception of end users, in terms of the working environment and the quality of support services necessary for the operations of the building assets as a business facility. The objective of measurement is to ensure the non-financial impact of the building facilities on occupants and visitors is aligned with core business policies and users' expectations. User and client considerations affect

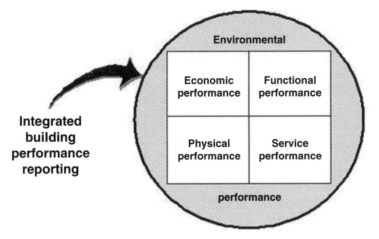

Figure 8.4 Integrated building performance reporting.

decision-making during building delivery as well as during occupancy. A broader definition of market needs analysis enables this feedback loop to address quality issues during the reuse and recycling phase of BPE.

Environmental metrics is concerned with the wider role of operational buildings and their impact on facilities users, the community and the ecological environment. Measures are likely to involve monitoring against prescribed sustainability targets at project/ state/national/international levels. Increasingly, these measures are expected to include the impact on the environment, and ecologically sustainable development considerations.

The premise is that any integrated building performance reporting must incorporate the five facets of measurement in order to obtain a balanced view of the contribution of building assets as an operating resource, as illustrated in Figure 8.4. The development of specific measures or KPIs for the five facets of performance is subject to a number of considerations and influences. A balanced framework of performance measures ensures that possible bias is mitigated (Then and Tan, 2004; Then and Clowes, 2004).

Figure 8.5 illustrates the components of a conceptual framework for measuring facilities management performance in buildings. The general push towards performance management is driven by well-founded principles of effective and efficient use of business resources (Audit Commission, UK, 2000; Eccles, 1991). Corporate management drives for continuous improvement have used external comparative analysis as a basis for validating internal performance. The nature of the business will determine the criticality of the operational building to the core business, in terms of their required attributes (e.g. location, image, etc.). Many corporations have been attracted by the opportunity to realign the balance sheet by total outsourcing – assets, liabilities and delivery. This trend has implications for measuring operational performance in buildings managed by external service providers.

Perspectives of performance can be expected to vary within the same organization, as well as between different organizations. Measurement of building performance is likely to vary according to the stakeholders' perspective and the selected time horizon. Although the outcomes may vary, the process of measurement is likely to support similar themes.

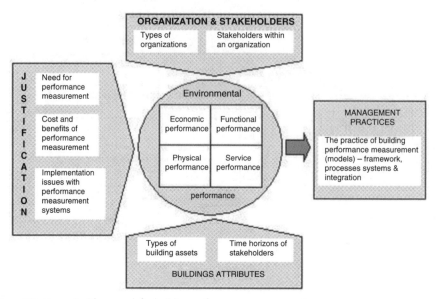

Figure 8.5 Conceptual framework for building performance measurement.

The resulting programme of building adaptation and reuse performance measurement in an organizational setting should represent the outcome of the preceding evaluations.

8.6   Conclusions

The quality of a building adaptation and reuse performance measurement programme is subject to the proper definition and organization of KPIs to provide relevant and reliable information for management decisions and actions. An unstructured and haphazard selection of KPIs is likely to lead to a waste of time and effort in data collection, and incomplete or misleading performance information. This chapter proposes a structured and logical framework for the development and selection of key performance measures. It provides a rational and robust methodology for the organization of the KPIs selected, and justification for data requirements. It also considers influencing factors in the selection of KPIs is and the potential bias arising from them. It opens the doors for further development of concepts relating to evaluation of building performance in the final phase of building performance evaluation, and the implementation of best practices. It is based on a balanced approach, recognition of the different management levels required for performance reporting, and the impact of external, influencing factors.

Acknowledgements

The author would like to acknowledge the following research collaborators, Mr Teng-hee Tan, Manager, Research and Development in Asset Management, Building Division,

Queensland Government, and Mr Andrew Clowes, Associate Director, Strategic Technology, Jones Lang Lasalle, Asia Pacific, and The Hong Kong Polytechnic University for funding the research.

References

Apgar, I.V. (1993). Uncovering your hidden occupancy costs. *Harvard Business Review.* May–June, pp. 124–136.

Asson, T. (2002). Real Estate Partnership: A New Approach to Corporate Real Estate Outsourcing. *Journal of Corporate Real Estate*, **4**, No. 4, 327–333.

Audit Commission (2000). *On Target – The Practice of Performance Indicators.* Audit Commission UK.

Becker, F. (1990). *The Total Workplace: Facilities Management and the Elastic Organisation.* New York: Van Nostrand Reinhold. Chapter 14.

Bon, R., McMahan, J. and Carder, P. (1994). Property Performance Measurement: From Theory to Management Practice. *Facilities*, **12**, No. 12, pp.

Damen, T. and Quah, L.K. (1998). Improving the Art and Science of Condition-based Building Maintenance. *Proceedings of CIB W070 Symposium on Management, Maintenance and Modernisation of Building Facilities – The Way Ahead into the Millennium* (L.K. Quah, ed.) pp. 141–148.

DeJonge, H. and Gray, J. (1996). The Real Estate Norm (REN). In *Building Evaluation Techniques* (G. Baird, N. Gray, D. Kernoham and McIndoe, eds), pp. 69–76. New York: McGraw-Hill.

Drake, A. (2002). Moving forward: Beyond Cost Per Square Foot – the Other Critical Success Factors in Workplace Change Projects. *Journal of Corporate Real Estate*, **4**, No. 2, 160–168.

Eccles, R.G. (1991). The Performance Management Manifesto. *Harvard Business Review.* January–February, pp. 131–137.

Hornec, S.M. (1993). *Vital Signs.* New York: AMACOM.

Kaplan, R. and Norton, D. (1996). *The Balanced Scorecard: Translating Strategy Into Action.* Boston, USA: Harvard Business School Press.

Leaman, A. and Bordass, W. (1999). Productivity in Buildings: The 'Killer' Variables. *Building Research & Information*, **27**, No. 1, 4–19.

Liebfried, K. and McNair, C.J. (1992). *Benchmarking: A Tool for Continuous Improvement.* New York: Harper Collins.

McBlaine, R. and Moritz, D. (2002). Transformational Outsourcing: Delivering the Promise. *Journal of Corporate Real Estate*, **5**, No. 1, 57–65.

McDougall, G., Kelly, J., Hinks, J. and Bititci, U. (2002). A review of the Leading Performance Measurement Tools for Assessing Buildings. *Journal of Facilities Management*, **1**, No. 2, 142–153.

Morgan, A. (2002). *The Importance of Property on Corporate Performance.* Working Paper 6. http://www.occupier.org

O'Mara, M.A. (2002). The global corporate real estate function: Organisation, authority and influence. *Journal of Corporate Real Estate*, **4**, No. 4, 334–347.

Preiser, W.F.E. (2001). The Evolution of Post-Occupancy Evaluation: Toward Building Performance and Universal Design Evaluation. Federal Facilities Council, 2001. *Learning from Our Buildings: A Summary of the Practice of Post-Occupancy Evaluation.* Washington, DC: National Academy Press.

Price, I. (2000). From Outputs to Outcomes: FM and the Language of Business. *Proceedings of CIB W070 Brisbane Symposium: providing facilities solutions to business challenges – moving towards integrated resources management. CIB Publication No. 235*, pp. 63–74.

Preiser, W.F.E. and Schramm, U. (1997). Building performance evaluation. In *Time-Saver Standards: Architectural Design Data* (D. Watson et al., eds), pp. 233–238. McGraw-Hill.

Szigeti, F. and Davies, G. (2001). Matching People and their Facilities: using the ASTM/ANSI standards on whole building functionality and serviceability. *Proceedings of CIB World Building Congress,* Wellington, New Zealand. Paper CLI 16.

Tan, T.H., Then, D.S.S. and Barton, R. (2000). Public Sector Asset Performance – Concepts and Implementation. *Proceedings of the CIBW70 Brisbane Symposium: Providing Facilities Solutions to Business Challenges – Moving towards Integrated Resources Management.* CIB Publication No. 235, pp. 17–29.

Then, D.S.S. (1994). Facilities Management – The Relationship between Business and Property. *Proceedings of Joint EuroFM/IFMA Conference – Facility Management European Opportunities,* pp. 253–262.

Then, D.S.S. (2002). Integration of Facilities Provision and Facilities Support Services Provision – A Management Process Model. *Proceedings of the 2002 Glasgow CIBW070 Global Symposium – Exploring the Global Knowledge Base on Asset Maintenance Management, Workplaces and Facilities Management.* CIB Publication No. 277, pp. 38–52.

Then, D.S.S. and Clowes, A. (2004). Stakeholders' Perspective in Defining Asset Performance. Under review for *CIB2004 Congress,* Toronto, May 2–7, 2004

Then, D.S.S. and Tan, T.H. (2004). Assessing Building Performance – An Integrated Model. Under review for *CIB2004 Congress,* Toronto, May 2–7, 2004.

Varcoe, B. (2002). The Performance Measurement of Corporate Real Estate Portfolio Management. *Journal of Facilities Management,* **1**, No. 2, pp. 117–130.

Vischer, J.C. (1996). *Workspace Strategies: Environment as a Tool for Work.* Chapman & Hall.

PART THREE

Case Studies

9

Benchmarking the 'sustainability' of a building project

Susan Roaf

Editorial comment

The buildings of the twenty-first century will have to perform radically better than those of the late twentieth century when, issues of climate change, fossil fuel depletion and security were not on the radar screen of building clients and users, engineers, architects and facilities managers. Building performance evaluation (BPE) routinely applied at every stage of a building's life, is a core tool in the delivery of performance improvements, and thus in the change needed to incorporate issues of sustainability into design and management of the building stock globally. Because buildings are the single largest source of terrestrial and atmospheric pollution, BPE may well become a key strategic tool in reducing environmental, social and economic impacts of buildings, and hence a key survival strategy for our society. This chapter outlines how BPE can incorporate sustainability indicators and benchmarks into the building process, and why it is important to do so.

9.1   Introduction: the problems are known

Buildings and cities are the most significant cause of environmental damage affecting our planet today. Major impacts on the environment include climate change, resource depletion, waste production, air, land, water and transport pollution from buildings, and the ensuing social deterioration that is typically associated with poor environmental conditions (Roaf et al., 2004a).

Since the early 1970s there has been growing concern about the probability of humankind, not only consuming more resources than the planet can provide (Goldsmith et al., 1972), but also polluting the planet's ecosystems to a point beyond which they can

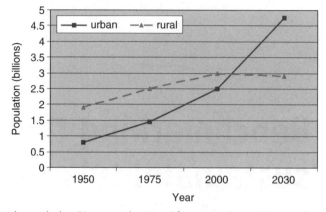

Figure 9.1 Growth over the last 50 years and projected future growth over 30 years of global urban and rural populations (http://www.jhuccp.org/pr/urbanpre.shtml).

no longer repair themselves (Meadows, 1972). Reinforcing these early concerns came the oil crisis of the 1970s, the emerging problems of ozone depletion and climate change in the 1980s, and the dawning realization that, far from being part of the solution, buildings are a large part of the problem (Roaf, 2004b). The scale of these problems is being exacerbated by the rapidly rising global population, and increasing consumer expectations.

Recent studies have shown that in the mid 1980s, the global population, crossed over the 'warning zone', where humans are beginning to exceed the carrying capacity of earth (Chambers et al., 2000). Indicators of this growing 'capacity' crisis include increasing water shortages in regions as diverse as Florida and Colorado and the Middle East, as well as events triggered by issues of energy insecurity caused by growing fuel shortages, terrorism, oil and gas price increases and infra-structural failures. That cities around the world from North and Latin America to Asia, the Middle East, India and Europe blacked out in 2003 provided a wake-up call for governments, city authorities and building owners that worse is to come, and such warnings will be ignored at the peril of all.

This chapter explains how office buildings are responsible for unsustainable resources consumption, and suggests how the building performance evaluation (BPE) approach might make commercial building projects more ecologically viable. The most potentially catastrophic global challenges we face today relate to the use and impacts of fossil fuels. Global climate change is linked to increasing levels of greenhouse gas emissions (IPCC, UKCIP and the Hadley Centre). Buildings alone are responsible for around 50 per cent of these anthropogenic emissions in developed countries such as the UK and the USA. BPE studies have demonstrated how much energy is used in buildings of different types since the late 1950s. In the UK, in 1990, the work of Bordass and Leaman highlighted the trend towards increasing levels of energy consumption, and ensuing carbon dioxide emissions in the office building sector (BRE, 1990).

Significant scientific work has gone into developing scenario-based studies of how much greenhouse gas emissions must be reduced globally in order to avoid the risk of destabilizing the climate. The UK Royal Commission on Environmental Pollution (RCEP, 2000) proposed realizable targets for the control of human-induced climate change in response to estimates of CO_2 emitted from the burning of fossil fuels that would lead to

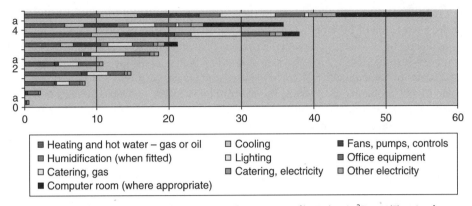

Figure 9.2 Office building types and their annual carbon emissions [*kgCarbon/m²TreatedFloorArea*] (**b** typical/**a** good practice), with extra category added for zero emission buildings: **0a** Zero emissions; **0b** Solar current best practice; **1** Naturally ventilated cellular; **2** Naturally ventilated open plan; **3** Air-conditioned standard; **4** Air-conditioned prestige[1] (diagram modified from Bordass, 1990).

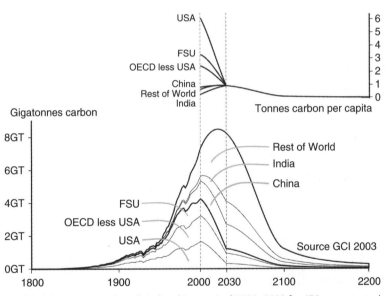

Figure 9.3 Global (six regions as shown) Carbon 'Contraction' 2000–2100 for 450 ppmv atmospheric concentration with 'Convergence' to equal per capita shares globally by 2030.
Source: Aubrey Meyer (see: http://www.gci.org.uk/)

[1] This diagram was based on one published in *ECON 19* by Bill Bordass, and unscientifically modified to include the category 0, by Isaac Meir and the author, to indicate how the low energy buildings of the future, powered by integrated renewable energies, may well perform.

high risks of catastrophic alterations to the climate. For the UK, this was calculated to require a reduction of 60% of CO_2 emissions by 2050 and 80% by 2100, relative to 1997 levels. This is based on the contraction and convergence principle, proposed by the Global Commons Institute, enshrining the idea that every human is entitled to release into the atmosphere the same quantity of greenhouse gases (Meyer, 2000).

The global problems are known, the macro-targets exist, and for key sectors of the building industry it is vital to integrate these pressing issues of 'sustainability' into the way in which we manage the design, use and disposal of our buildings.

9.2   The role of BPE in making buildings sustainable

Traditionally, the boundary conditions of building performance evaluation have been the four walls of the building. However this must change if buildings are to be made more accountable for their impacts. Buildings must be accounted for as part of their environmental context at local, regional and global levels if impacts are to be measured, managed and reduced.

There are two mechanisms through which this could be effected, using current building industry auditing processes:

1. The sustainability statement or evaluation. Each building undergoes a sustainability evaluation, measured against a standard checklist of recognized sustainability issues, indicators and benchmarks such as those developed by the State of Minnesota (Minnesota, 2000a) for the predesign, design, construction and occupation stages of the project (Minnesota, 2000b).
2. Building performance evaluation. The BPE framework (Preiser, 2001) can incorporate some core indicators and benchmarks of sustainability into the performance evaluation remit of the BPE effector (E), incorporated into a single building project at the pre-design (G and C), design (O and P), construction (O and P), and occupation (O and P) stages of the building's life (see Figure 9.4).

As BPE is being increasingly incorporated into best practice management of the design and operation of buildings by owners, designers and facilities managers, it would seem sensible to eliminate the need for two performance evaluations and to concentrate on refining and improving BPE.

9.3   The rationale for incorporating issues of sustainability

The problems we face pose such a significant threat to our way of life and to the planet, that a number of drivers for change are already emerging that may effect radical changes in the building industry. These are presented below.

9.3.1   Legislation

The European Union has perhaps the most advanced policies for 'green' governance in the world, manifested in a rolling programme of social, economic, resource and environmental impact policies that are resulting in wide-ranging legislation that is beginning to affect a number of sectors. This includes the European Building Directive which, from 2007 onwards, requires every house to have a performance certificate issued at point of sale, and every public building in Europe to have an annually renewed Performance Certificate, much like an automobile must have, displayed openly in a public place in the building (EPB Directive 2002/91/EC). Incorporating issues of sustainability could strengthen the case for making the BPE approach statutory in some countries.

9.3.2   Corporate social responsibility

While regulation is forcing the pace of change in Europe, the USA and elsewhere, the 'greening' of businesses is also being driven by the rapid uptake of the concept of corporate social responsibility (CSR). CSR is a movement based on the growing realization of the significance of the 'virtuous circles' associated with improved building performance and the significant market advantages of '*do well by doing good*' (Hampton, 2004). As Kofi Annan, Secretary General of the United Nations stated: 'CSR will give a human face to the global market' and 'embrace, support and enact a set of core values in the areas of human rights, labour standards and environmental practices'. (www.unglobalcompact.org)

CSR thinking encourages companies to combine profitability and sustainability by making this a cornerstone of the way they operate relative to four streams:

- Philanthropy and community involvement
- Socially responsible investment
- Social auditing
- Reduction of environmental impacts of business

As the mindset of commercial building owners and managers changes, so all those involved in the related process of enhancing building performance will actually become involved in the greening of buildings, from board members themselves, to government agencies, planners, architects, engineers, surveyors, estate agents, facilities managers, building owners and occupants and the communities around them.

The starting point for such changes in the context of BPE (see Figure 9.4) is the client's 'big picture' goals (G); for example, 'I want the greenest building in Colorado.' The need then arises to understand what the key issues, and ensuing criteria, for achieving a green building in Colorado are (C), and to set appropriate indicators (O) and benchmarks (P), and targets of performance in relation to those local and regional sustainability issues (C), for example, water use. These targets will in turn reflect the shade of green required (is it to be a dark, medium or light green or grey building?) in a proactive process of performance improvement.

The challenge is then to translate these performance indicators into useable benchmarks for a particular building (P), for example, water use in litres per capita per annum, and to establish a methodology with which to measure easily and regularly the performance of the building in relation to those indicators. The actual performance of the building can then be evaluated systematically over its lifetime, and compared to the more general benchmarks available, to establish the extent to which the client's 'green goals' and aspirations have been met.

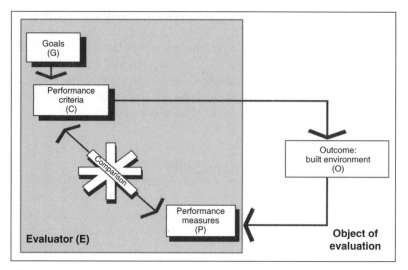

Figure 9.4   Performance concept/evaluation system.
Source: Preiser (1991, 2001).

9.4   The client's choices

Many decisions in the early stages of a project flow from the client's goals (G) that will rest, initially, on the choice between the two following options put forward by Hyams (2004):

● The first option proposes minor technical changes to current practice and looks to future research to find solutions to longer-term problems, without accepting the need for any fundamental changes to the underlying nature of development. It does not apply the precautionary principle to environmental issues or risk evaluations, and nor does it suggest any redesign of operational activities.
● The second option is radical, responding to the warning signs seen in, say, pollution, fossil fuel depletion and climate change arising from current practices in the traditional economy and its exploitation of natural resources. It proposes a new appreciation of nature and human life that will change the way that buildings and developments are conceived. The goals of clients choosing this option become radical, and indicators and benchmarks set demanding environmental performance targets, which result in significant reductions in impacts.

It is important to understand where the client stands in relation to the two options outlined above, as well as what they are willing to pay for. Without an understanding of which sustainability goals are required, design guidelines and benchmarks are ineffectual. Adding on a few environmentally friendly objectives can do little to mitigate the effects of an inherently unsustainable project. Thinking about sustainability is fundamental to any project and it starts in the period before formal briefing commences – that is, Phase 1 (Strategic Planning) of BPE. If the client wishes to make all of its activities fundamentally environmentally sustainable, the buildings it commissions and operates will reflect its core policies; and if not, then they probably won't (Hyams, 2004).

In order to achieve a relatively green building, issues and indicators (C and O) and benchmarks for the performance of a building (P) must be identified that are:

1. Linked to fundamental concepts of sustainable development such as those of:
   - 'equity' – the equalizing of resource consumption around the world and
   - 'futurity' – the need to ensure that future generations are left with sufficient resources to live well, and
   - 'climate stabilization' – the need to claw back our emissions of greenhouse gases to the extent that the climate will not be irreversibly changed by our profligate burning of fossil fuels.
2. Agreed on between clients and all building stakeholders at the initial predesign stage, and consistently applied at each subsequent stage of the design, construction and occupation of the building.

9.5   What are the key issues of sustainability ?

Core sustainability and survival issues relate to:

- Resource depletion 'not exceeding the capacity of our planet to support our life styles'. This includes not only physical resources such as fossil fuels and materials, but also the human resource, ensuring that quality of life of individuals and communities alike is optimized.
- Pollution not 'fouling the nest'. This issue includes the need to reduce waste, air and water pollution, and ozone depleting and greenhouse gas emissions by controlling, for instance, building and related transport emissions.

A number of excellent front end 'sustainability guidelines' exist, including the Minnesota Sustainable Guidelines (2003), the UK SEEDA Sustainability Checklist (2002), and published checklists such as the Green Guide to the Architect's Job Book (Halliday, 2000). More work needs to be done on tying more broadly-defined targets to the design and operational benchmarks of performance included in such guidelines.

While such standardized indicator and benchmark sets have the advantage that they enable the comparison of one building to another, they may not reflect the actual performance of an individual building. Robust indicators and benchmarks that are locally referential, as well as regionally and globally relevant, must be selected.

9.6   What are indicators and benchmarks?

An indicator is a parameter, or a value derived from parameters, that describes the state of the environment and its impact on human beings, ecosystems and materials; the pressures on the environment; and the driving forces, and the responses steering that system. An indicator has gone through a selection and/or aggregation process to enable it to steer action (http://glossary.eea.eu.int/EEAGlossary/E/environmental_indicator).

A benchmark is defined as a measurable variable used as a baseline or reference in evaluating the performance of an organization (or environmental indicator). Benchmarks may be drawn from internal experience or that of other organizations, or from legal requirements, and are often used to gauge changes in performance over time (http://www.snw.org.uk/enwweb/business/busenvglos.htm).

Simply put, the indicator (O) is a yardstick of an issue (C), such as water, energy or waste, and the benchmark (P) is the point at which a particular building or a set of buildings occurs on that yardstick, giving a measure of relative performance against other measured buildings.

The initial step is to choose appropriate criteria to form an 'indicator set' for a project, which relates the building's performance to the nature, requirements and resources of the local environment, culture and economy, as well as to business goals. Efficiencies will be built into the process and the product by understanding local conditions and resources. Where large buildings are involved, it may be sensible to involve a sustainability expert to help develop the indicator and benchmark sets (Graham, 2003). Alternatively, many local government councils in the UK and the USA are employing their own specialists in this field.

9.7   What issues should be included in an indicator set?

For a full set of sustainability issues, indicators and benchmarks see Minnesota (2000), SEEDA (2003) and Roaf et al. (2004a). Many of the related performance questions may be asked as part of good practice for building managers. But to understand them in a context of sustainability, these questions need to be related to the capacity of the ecosystem to support them. For example:

- How much energy does it use? What are the 2050 targets it should work towards?
- How much CO_2 does it produce? What is its fair earth share of emissions?
- How much water is used per capita? How much can the region supply per capita?
- How much waste is produced from it? What levels can the local ecosystem support over time and what are the fiscal penalties related to the waste levels generated?
- What are the emissions related to the transport miles it generates? What are the transport related targets in the building's travel plan?
- How happy are its occupants? And why? How much happier can they be made and how?
- How much pollution does it generate? How do these levels relate to legislative targets over time?
- How does it affect biodiversity in the region? How do these impacts relate to biodiversity strategies in the region?
- How sustainable, durable, replicable, maintainable are the materials it is made of?
- How can the resilience of the materials of the building be improved?
- How durable are the functions of the building?
- How adaptable is the building for other functions over time?

To place the performance of an individual building in the context of regional and global benchmarks, it can be compared to (Bordass, 2002, see also Chapter 7):

- Absolute benchmarks: a benchmark measure that represents a statistically derived national or regional average for the performance of buildings in general, or particular building types, such as offices, against that indicator.
- Relative benchmarks: a benchmark that provides a more closely related, comparative measure for a similar, clearly defined or 'reference' building type, such as a standard or prestige office building.

● Tailored benchmarks: a benchmark derived from reference buildings that is 'tailored' to optimize the usefulness of the diagnostic potential that can be gained from the comparison, and to provide common bases for comparison, such as occupation densities, hours of use, and type and amount of equipment.

9.8   Conclusions

The process of embedding the evaluation of building sustainability within the BPE process is becoming increasingly important as we are made aware of the challenges ahead, including those of fuel insecurity, terrorism, climate change, and resource depletion. In addition, the social and economic costs of powering commercial buildings are increasing as fossil fuels become scarcer and more expensive.

A key challenge of the BPE approach is not only to incorporate such issues into the BPE process, but also to ensure that businesses are aware of the social, economic and environmental advantages of doing so, and of the 'virtuous circles' of investment and reward that are the real business benefits of a truly sustainable building. The carrot often works better on the trading – or the boardroom – floor than the stick.

References

Accountability (2003). See: http://www.accountability.org.uk

Bordass, B. (1990). Appropriate methods and technologies for new build and refurbishment: offices, Global Responsibilities of Architects, RIBA publications, pp. 15–17.

Bordass, W. (2001). *Flying blind: everything you always wanted to know about energy in commercial buildings but were afraid to ask.* Association for the Conservation of Energy and Energy Efficiency Advice Services for Oxfordshire. London, October.

BRE (1990). Energy efficiency in commercial and public sector offices. *Best Practice Energy Programme Energy Consumption Guide 19.* UK Department of the Environment/Building Research Energy Conservation Support Unit, Building Research Establishment, UK. See: http://www.targ.co.uk/other/guide19.pdf

Chambers, N., Simmons, N. and Wackernagel, M. (2000). *Sharing Nature's Interest: Ecological Footprints as an Indicator of Sustainability.* Earthscan Publications Ltd, London, p. 51. Forum for the Future (2003). For a fuller discussion of CSR see: http://www.forumforthefuture.org.uk/publications/default.asp?pubid=50

EPB Directive 2002/91/EC was formally adopted by the Council of Ministers on the 16 December, 2002, and published in the EU Official Journal on 4 January 2003. It is expected to result in savings of 45 million tonnes of CO_2 by 2010. For an excellent discussion site on the implication of the Directive for UK buildings see: http://www.europrosper.org

Goldsmith, E., Allen, R., Allenby, M., Davull, J. and Lawrence, S. (1972). *A Blueprint for Survival.* Tom Stacy Ltd, pp.106–107.

Graham, P. (2003). *Building Ecology: First Principles for a Sustainable Environment.* Blackwell Publishing.

Hadley Centre in the UK deals with climate modeling and a full overview, with publications, can be seen on: http://www.met-office.gov.uk/research/hadleycentre

Halliday, S. (2000). *Green Guide to the Architect's Job Book.* RIBA Publications.

Hampton, D. (2004). The Client's Perspective – CSR. *Closing the Loop: Benchmarks for Sustainable Buildings.* In S. Roaf, A. Horsley and R. Gupta, RIBA Publications: in press.

Hulme, M., Jenkins, G. J., Lu, X., Turnpenny, J. R., Mitchell, T. D., Jones, R. G., Lowe, J., Murphy, J. M., Hassell, D., Boorman, P., McDonald, R. and Hill, S. (2002). *Climate change scenarios for the United Kingdom: the UKCIP02 scientific report.* Tyndall Centre for Climate Change Research, School of Environmental Sciences, University of East Anglia, Norwich, UK.

Hyams, D. (2001). *Briefing.* RIBA Publications.

Hyams, D. (2004). Briefing. *Closing the Loop: Benchmarks for Sustainable Buildings.* In S. Roaf, A. Horsley and R. Gupta (eds). RIBA Publications: in press.

IPCC. The International Panel on Climate Change is the lead international organization on climate change and a full review of their work can be found on: http://www.ipcc.ch

IPCC (2001). *Climate change 2001: the scientific basis: contribution of working group I to the third assessment report of the intergovernmental panel on climate change.* Cambridge University Press.

Meadows, D.H. (1972). *The Limits to Growth: A Report for the Club of Rome's Project on the Predicament of Mankind.* St. Martin's Press.

Meyer, A. (2000). *Contraction and Convergence: the global solution to climate change.* Green Books for the Schumacher Society.

Minnesota (2000a). See the following site for an overview of the stages of the process of integrating sustainable design issues into a building project at the predesign, design, construction and occupancy phases: http://www.sustainabledesignguide.umn.edu/MSDG/text/phases.pdf

Minnesota (2000b). See the following websites for a full overview of the design guide criteria, indicators and benchmarks: http://www.sustainabledesignguide.umn.edu/MSDG/resources5_2.html and http://www.csbr.umn.edu/B3/index.html

Minnesota Sustainable Buildings Guidelines (2003). http://www.csbr.umn.edu/b3/

Preiser, W.F.E. (1991). Design intervention and the challenge of change. In *Design intervention: toward a more humane architecture* (W.F.E. Preiser and J.C. Vischer, eds). Van Nostrand Reinhold.

Preiser, W.F.E. (2001). The evolution of post-occupancy evaluation: toward building performance and universal design evaluation. In *Federal Facilities Council. Learning from our buildings – a state-of-the-practice summary of post-occupancy evaluation.* National Academy Press.

RCEP (2000). *Energy – the changing climate: summary of the Royal Commission on Environmental Pollution's Report.* London: HMSO.

Roaf, S., Horsley, A. and Gupta, R. (2004a). *Closing the Loop: Benchmarks for Sustainable Buildings.* RIBA Publications.

Roaf, S., Crichton, D. and Nicol, F. (2004b). *Adapting Buildings and Cities for Climate Change.* Architectural Press.

SEEDA (2002). See: www.sustainability-checklist.co.uk

UKCIP. For a full online overview of the UK Climate Impacts Programme online see: http://www.ukcip.org.uk/

Authoritative indicator sets (beware of hidden agendas)

Global

The World Health Organization (http://www.who.int/en/).

The United Nations Programme for the Environment (http://www.unep.org/).

The United Nations Sustainability Indicators (http://www.un.org/esa/sustdev/natlinfo/indicators/isd.htm).

The International Standards Organization (http://www.iso.ch/iso/en/ISOOnline.openerpage). This body has members from 147 countries, responsible for 13 700 ISO standards and provides a bridge between the public and industry, and ensures the rectitude of the commonly agreed standards used in business and government. It is here that the influential standards for environmental monitoring and management have been developed in ISO: 14001 (http://www.iso.ch/iso/en/iso9000-14000/iso14000/iso14000index.html).

European

www.eea.eu.int/
http://europa.eu.int/comm/enterprise/environment/index.htm

Websites for Laws

Global

For a list of United Nations Treaties, including the International Convention on Climate Change see: http://untreaty.un.org/
Issues dealt with by UN programmes can be seen on:
http://esa.un.org/subindex/pgViewTerms.asp?alphaCode=a

European

EU Laws are to be found on links from:
http://europa.eu.int/comm/environment/legis_en.htm

10

Introducing the ASTM facilities evaluation methodology

Françoise Szigeti, Gerald Davis, and David Hammond

Editorial comment

The American Society for Testing and Materials (ASTM) methodology and tools provide a way both to define requirements, and to perform building performance evaluation in a structured, calibrated manner so that results can be compared at each phase of the BPE (see Chapters 1 and 2). This chapter explains how the ASTM/ANSI standards for whole building functionality and serviceability (ASTM standards) help decision-makers and stakeholders acquire information about their real estate assets to enable them to make suitable choices, adjudicate trade-offs, set priorities, and allocate budgets throughout the complete life cycle of such assets. The chapter describes the methodology and tools that have been standardized (ASTM, 2000), and includes an example to demonstrate how information captured in this manner can be linked to virtual building models and other information such as building condition and service life, space utilization, etc. The chapter concludes with a review of current trends, and points to the increasing need to break down traditional silos of information and to make connections throughout the different phases of the life cycle of a facility.

10.1   Overview

The serviceability tools and methods (ST&M) approach includes ASTM standard methodology and scales (Davis et al., 1993). The standards are used to define building requirements – for occupants, managers and owners – and to measure how well a facility meets those requirements. They are based on matching scales to compare what is required with what is provided. This approach has been shown to ease communication among stakeholders, encourage collaboration, and provide essential content for portfolio and asset management, for strategic plans, for statements of requirements (SoR), and for budget priorities. It is a tool for use in the feedback loops of several phases of building performance evaluation, including Phase 1, strategic planning; Phase 2, programming; Phase 4, occupancy; and Phase 6, adaptive reuse and recycling.

The ASTM standards comprise a set of scales that form a methodology to:

- link real estate decisions to the mission of the organization or to the objectives of decision-makers
- create and use calibrated scales to define requirements (demand) on a wide range of topics and to assess the capability of facilities (supply) to meet these stated requirements at a matching level
- define demand
- rate supply
- match demand to supply, and
- analyse the gaps.

The scales are calibrated and structured to provide comparable results fast, reliably, and at a relatively low cost. They are not intended for detailed programming or in-depth technical investigations. The scales are explicit, comprehensive, transparent, and easy to use with little training. The results are expressed as qualitative descriptions and quantitative results that are auditable and congruent with ISO 9000.

The scales capture customer-defined requirements and match them to indicators of capability expressed as performance-based requirements. The indicators of capability can be used to assess how well a proposed design, or an existing facility (occupied, or proposed to be leased or bought), meets the functional requirements specified by organizational units and facility occupants.

Information about functionality can be combined with other building data in a comprehensive 'suitability stamp' that gives decision-makers an overview of their real estate assets in relation to their requirements. It shows graphically which facilities are at risk and require urgent action, as well as issues that require attention. The ability to use computerized databases and virtual building models will, in time, allow all stakeholders access to the same information throughout the life cycle management process. The example included in this chapter illustrates how the ASTM standards method and scales are used for a specific project.

10.2   Methodology and tools

10.2.1   Using the ASTM standards with the serviceability tools and methods approach

The ST&M approach (Davis et al., 1993) provides formats for describing organizational needs, function-based tools for estimating floor area requirements, and other tools necessary to identify requirements. The relationship of a statement of requirements to goals and objectives of clients is part of the overall performance systems model. This offers a framework for both regulatory and non-regulatory performance-based building (Meacham et al., 2002; Hattis and Becker, 2001; Gib   1982), of which BPE is one example (Preiser and Schramm, 1997).

Many of the techniques and t   in b⁊ ⁊   pture this information have been catalogued by researchers and pr𝔞   ′ersity of Victoria in Wellington,

New Zealand (Baird et al., 1996), including the ST&M approach described in this chapter. Other tools are described in Chapters 11, 14 and 15 of this book.

At the heart of the ST&M approach is the process of working with occupant groups and stakeholders. This process of communication between the providers of services and products (in-house and external) and other stakeholders (in particular the occupants) of valuing their input, and of being seen to be responsive, can be as important as the outcome itself, and will often determine the acceptability of the results. This is where satisfaction, perception and quality overlap. The ST&M approach includes:

1. a profile of functionality requirements, usually displayed as a graphic bar chart and described in a written format as a text profile of required functional elements;
2. a profile of facility serviceability, which can be displayed as a graphic bar chart, and described in written format as a text profile of indicators of capability;
3. a match between any two profiles and comparisons with up to three profiles;
4. a 'gap analysis', including determining priorities and concerns for presentation to senior management in summary bar charts and tables, a table of actions required, and a suitability stamp for each main building, facility, zone and variant;
5. text profiles for use in a statement of requirements, including equivalent indicators of capability;
6. a descriptive text about the organization, its mission, relevant strategic information, and other information about the project in a standard format;
7. quantity spreadsheet profiles and estimated space envelopes;
8. 'building loss features' (BLF) rating table and assessment of space utilization, including rentable versus programmable;
9. a footprint and layout guide, including workstation footprints and building geometry.

10.2.2   Why define functionality requirements and quality?

Quality is described in ISO 9000 as the totality of features and characteristics of a product or service that bear on its ability to satisfy stated and implied needs. Quality is also defined as fitness for purpose at a given cost; it is not absolute, but relative to the circumstances. It is the most appropriate result that can be obtained for the price one is willing to pay. In order to be able to evaluate and compare different results or offerings and verify whether requirements have been satisfied, these requirements must be stated as clearly as possible (Szigeti and Davis, 2002).

Assessing customer perception and satisfaction, and evaluating the quality of the performance delivered by a facility in support of customer requirements, are two complementary, but not identical, types of assessment. In a series of articles dealing with the ratings of health maintenance organizations, Consumer Reports makes the following point: 'Satisfaction measures are important. But, don't confuse them with measures of medical quality ...' (Consumer Reports, 2000). Measuring customer satisfaction is important and necessary, but it is equally important to measure the quality and performance of the services and products delivered separately, whether it be medical care or facilities and services for occupants and business owners.

Portfolio and asset management provide the link between business demands and real estate strategy. At the portfolio level, requirements for facilities are usually rolled into the strategic real estate plan in support of the business plan for the enterprise (Szigeti and Davis, 2001a; Szigeti, F. et al., 2004); requirements for facilities are normally included in a portfolio management strategy. An asset management plan for each facility includes the specific criteria to enable an evaluation of the quality of the performance of that facility. A statement of requirements, in one form or another, is part of the contractual documentation for each specific procurement. Many enterprises, public and private, typically review the project file during commissioning and hand-over, and note whether the project was completed within budget and on schedule (see Chapter 6 on commissioning). Some also assess how well each new or renovated facility meets the requirements of the business users who occupy it (see Chapter 7 on post-occupancy evaluation). Essential knowledge can be captured on what works well, what works best, and what should not be repeated in an institution's buildings.

10.2.3   Comparing functionality and serviceability

The ASTM standards provide a broad-brush, macro-level method, appropriate for strategic decision-making. They can be used to ascertain the level of functionality required of each facility on a wide range of topics, and to assess its physical features as indicative of its level of capability, that is, how serviceable it is in terms of each requirement. For each topic, the approach uses a pair of ASTM standard scales, one for demand to assess one 'functionality' topic, and one for supply to assess one 'serviceability' topic. This is shown in Figure 10.1.

10.2.4   The ASTM standard scales

Each ASTM standard scale includes two matched, multiple-choice questionnaires that are paired to deal with both demand (occupant requirements) and supply (serviceability of buildings). The concept of demand and supply, and of the two languages used by each, has long been used in finance and business, and is now widely accepted in the property management community (Ang et al., 2002; McGregor and Then, 1999).

The ASTM standard scales are designed as a bridge between functional programmes written in user language, and outline specifications and evaluations written in technical performance language. Although it is standardized, this approach has proven to be easily adaptable to the needs of a specific organization. For organizations with many facilities that house similar types of functions, ASTM standards capture a systematic and consistent record of the institutional memory of the organization. Their use speeds up the functional programming process, and provides comprehensive, systematic, objective ratings in a short time.

The scales in the ASTM standards cover more than 100 topics and 340 building features, each with levels of service calibrated from 0 to 9 (less to more). Not all scales need to be used on each project. Some organizations have used one scale to assess their whole portfolio; others have selected about 50 to be used for typical projects. The scales are illustrated in Figure 10.2.

For each set of ASTM standards, there are two multiple-choice questionnaires. One is used for setting occupant requirements for functionality and quality. This questionnaire describes customer needs (demand) in everyday language, and user responses are typically gathered during individual and group interviews.

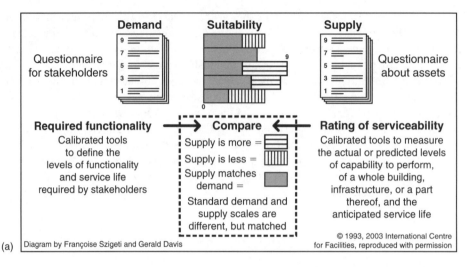

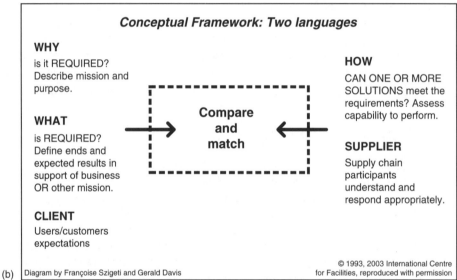

Figure 10.1 (a) Matching demand and supply to prepare a gap analysis; (b) Two languages: 'client/user' and 'supplier/provider'.

The other matching multiple-choice questionnaire is used to assess the serviceability of the building. This questionnaire allows people who are not technical experts to rate the capability of the facility (supply) based on descriptions in performance language; for example, as a first step toward an outline performance specification (see Chapter 2 on feedback loops in building performance evaluation).

The ASTM standards are project-independent. They can be used at any time and updated at any point during the life cycle of a facility. Scales have been developed to evaluate service life and condition of maintenance shops and service yards, added security, and codes and regulations. Some are still in the development stage (Davis et al., 2001; Davis et al., 2000a and 2000b).

Letters alongside a topic number and name indicate a topic selected for a specific profile. The left column, (of **F**s), indicates topics selected by a group providing facilities to the components of a large organization. The next column, (of **P**s), shows topics selected by a unit with a 'people' priority to attract and retain high quality staff in the changing labour market of the next decade or two.

A. GROUP AND INDIVIDUAL EFFECTIVENESS

A.1 Support for Office Work
P A.1.1   Photocopying and office printers
    A.1.2   Training rooms, general
    A.1.3   Training rooms for computer skills
    A.1.4   Interview rooms
F A.1.5   Storage and floor loading
F A.1.6   Shipping and receiving

A.2 Meetings and Group Effectiveness
F P A.2.1   Meeting and conference rooms
F P A.2.2   Informal meetings and interaction
F P A.2.3   Group layout and territory
P A.2.4   Group workrooms

A.3 Sound and Visual Environment
F P A.3.1   Privacy and speech intelligibility
F P A.3.2   Distraction and disturbance
F A.3.3   Vibration
P A.3.4   Lighting and glare
F P A.3.5   Adjustment of lighting by occupants
P A.3.6   Distant and outside views

A.4 Thermal Environment and Indoor Air
F P A.4.1   Temperature and humidity
F P A.4.2   Indoor air quality
F P A.4.3   Ventilation air (supply)
P A.4.4   Local adjustment by occupants
P A.4.5   Ventilation with operable windows

A.5 Typical Office Information Technology
F P A.5.1   Office computers and related equipment
F P A.5.2   Power at workplace
F A.5.3   Building power
F P A.5.4   Telecommunications core
F P A.5.5   Cable plant
F A.5.6   Cooling

A.6 Change and Churn by Occupants
F P A.6.1   Disruption due to physical change

F A.6.2   Illumination, HVAC and sprinklers
F P A.6.3   Minor changes to layout
    A.6.4   Partition wall relocations
    A.6.5   Lead time for facilities group

A.7 Layout and Building Features
P A.7.1   Influence of HVAC on layout
P A.7.2   Influence of sound and visual features on layout
    A.7.3   Influence of building loss features on space needs

A.8 Protection of Occupant Assets
F P A.8.1   Control of access from building public zone to occupant reception zone
F P A.8.2   Interior zones of security
    A.8.3   Vaults and secure rooms
F A.8.4   Security of cleaning service systems
    A.8.5   Security of maintenance service systems
    A.8.6   Security of renovations outside active hours
    A.8.7   Systems for secure garbage
F P A.8.8   Security of key and card control systems

A.9 Facility Protection
F A.9.1   Protection around building
F P A.9.2   Protection from unauthorized access to site and parking
P A.9.3   Protective surveillance of site
P A.9.4   Perimeter of building
    A.9.5   Public zone of building
P A.9.6   Facility protection services

A.10 Work Outside Normal Hours or Conditions
F P A.10.1   Operation outside normal hours
F A.10.2   Support after-hours
P A.10.3   Temporary loss of external services
    A.10.4   Continuity of work (during breakdowns)

Figure 10.2 List of topics – with several choices for different uses (Continued).

A.11 Image to Public and Occupants (E 1667)

F	A.11.1	Exterior appearance
F P	A.11.2	Public lobby of building
F P	A.11.3	Public spaces within building
F P	A.11.4	Appearance and spaciousness of office spaces
F P	A.11.5	Finishes and materials in office spaces
	A.11.6	Identity outside building
F P	A.11.7	Neighbourhood and site
	A.11.8	Historic significance

A.12 Amenities to Attract and Retain Staff

F P	A.12.1	Food
	A.12.2	Shops
P	A.12.3	Day care
F P	A.12.4	Exercise room
	A.12.5	Bicycle racks for staff
P	A.12.6	Seating away from work areas

A.13 Special Facilities and Technologies

F P	A.13.1	Group or shared conference centre
P	A.13.2	Video teleconference facilities
	A.13.3	Simultaneous translation
	A.13.4	Satellite and microwave links
F	A.13.5	Mainframe computer centre
	A.13.6	Telecommunications centre

A.14 Location, Access and Wayfinding

F P	A.14.1	Public transportation (urban sites)
	A.14.2	Staff visits to other offices
F P	A.14.3	Vehicular entry and parking
P	A.14.4	Wayfinding to building and lobby
F P	A.14.5	Capacity of internal movement systems
F P	A.14.6	Public circulation and wayfinding in building

B. THE PROPERTY AND ITS MANAGEMENT

B.1 Structure, Envelope and Grounds

F	B.1.1	Typical office floors
	B.1.2	External walls and projections
F	B.1.3	External windows and doors
	B.1.4	Roof
	B.1.5	Basement
F	B.1.6	Grounds

B.2 Manageability

	B.2.1	Reliability of external supply
	B.2.2	Anticipated remaining service life
	B.2.3	Ease of operation
	B.2.4	Ease of maintenance
	B.2.5	Ease of cleaning
	B.2.6	Janitors' facilities
F P	B.2.7	Energy consumption
F	B.2.8	Energy management and controls

B.3 Management of Operations and Maintenance

F	B.3.1	Strategy and programme for operations and maintenance
	B.3.2	Competence of in-house staff
F P	B.3.3	Occupant satisfaction
	B.3.4	Information on unit costs and consumption

B.4 Cleanliness

F P	B.4.1	Exterior and public areas
F P	B.4.2	Office areas (interior)
F P	B.4.3	Toilets and washrooms
F	B.4.4	Special cleaning
P	B.4.5	Waste disposal for building

Figure 10.2 (Continued).

10.2.5   Categories for action: summarizing results from a gap analysis

Once the functionality requirements profiles and the serviceability rating profiles are complete, they can be compared. A gap analysis is then conducted, which summarizes the topics for which there is a significant surplus or shortfall. Figure 10.3 shows both a comparison between functionality and serviceability profiles, and a gap analysis of significant discrepancies. When assessing a portfolio of properties, the information can be summarized for a

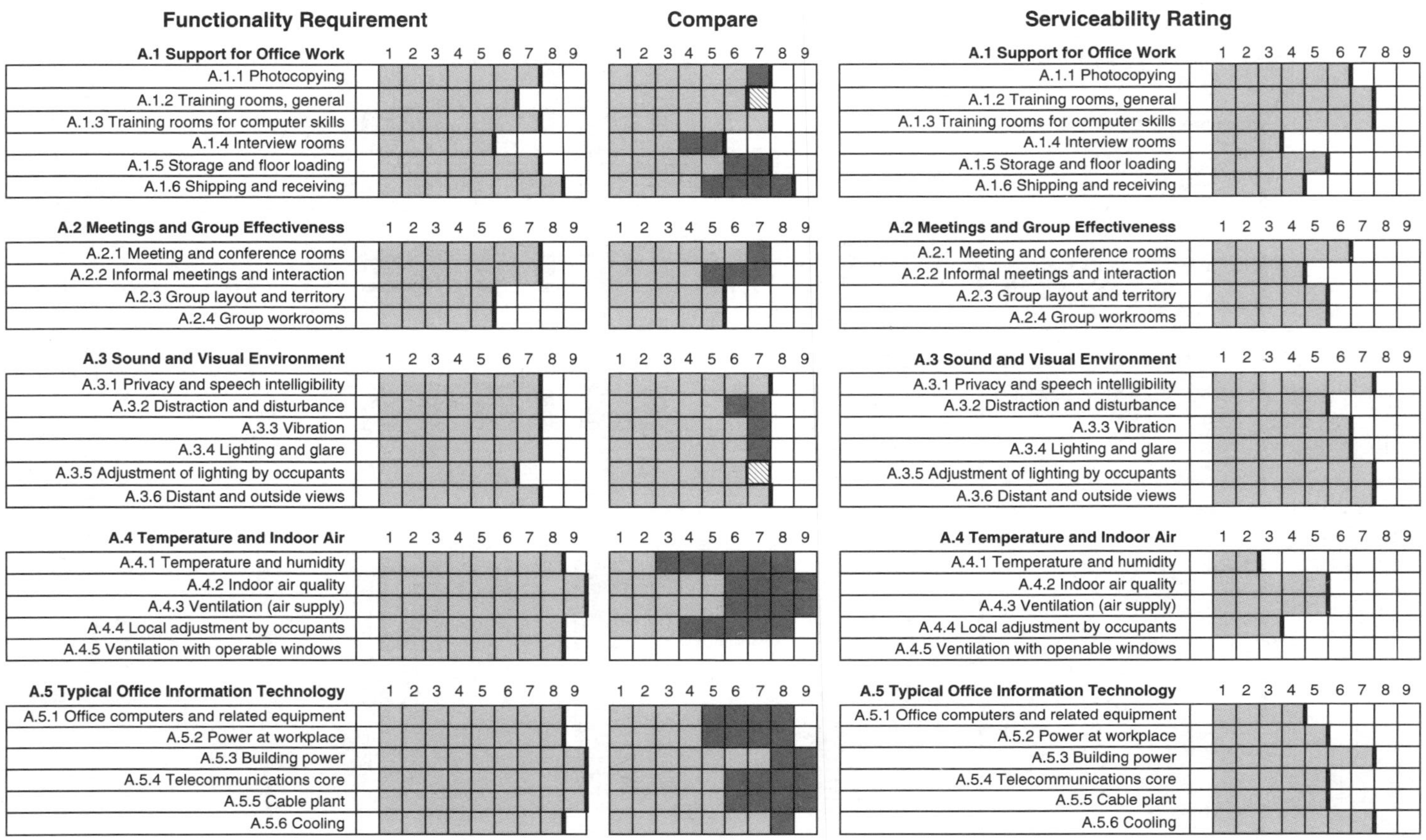

Figure 10.3 Sample comparison bar charts for gap analysis and significant discrepancies.

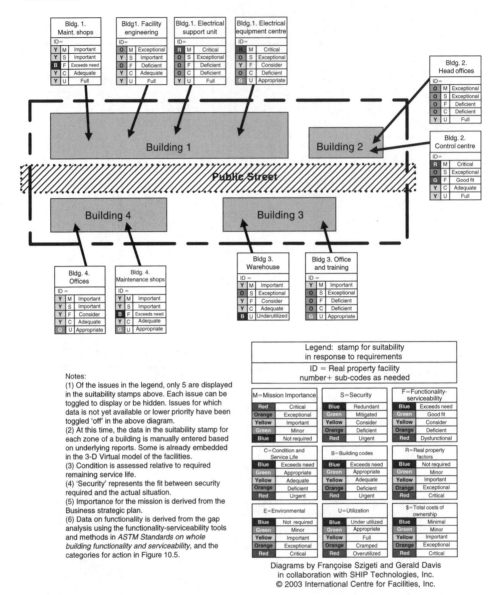

Figure 10.4 Suitability in response to requirements (site map with suitability stamps).

building or a geographic area. On the basis of these summaries, senior executives can be provided with a recommended course of action. For a portfolio of assets, this provides the ability to cluster facilities in categories for action and budgetary priorities, and to prepare suitability stamps (see Figures 10.4 and 10.5).

This methodology builds on a client's in-house experience and knowledge, so that the wheel is not reinvented for each building project. Results are based on explicit, documented

Count of topics considered	14	13	14	14	10	43	24
	Bldg 4 Maint. Shops	Bldg 1 Maint. Shops	Bldg 2 Control Centre	Bldg 3 Ware-house	Bldg 1 Electr. Equips Centre	Bldg 4 Offices	Bldg 1 Elect. Maint. Office
Topics lack information, so level cannot be set	0	2	0	0	0	12	0
Topics with sufficient info to be able to rate	**14**	**11**	**14**	**14**	**10**	**31**	**24**
Topics with significant problems of fit							
Serviceability does not meet minimum threshold level	0	0	0	1	2	5	7
Exceptionally important topic, at least 2 levels below requirement (and not in above count of missed threshold levels)	0	0	0	0	0	2	3
Important topic, and at least 3 levels below requirement (and not in above count of missed threshold levels)	0	0	0	1	0	5	4
Minor importance topic, 4 or more levels below or above requirement (and not in above count of missed threshold levels)	0	0	0	0	0	0	0
Count of topics with significant problems of fit	0	0	0	2	2	12	14
Per cent of topics **without** significant problems of fit. Formula is: $\dfrac{\text{Topics sufficient info} - \text{Topics problems of fit}}{\text{Topics sufficient info}}$	100%	100%	100%	86%	80%	61%	42%
Count of topics that exceed needs by more than one level	7	5	1				
Per cent of topics that exceed need by more than 1 level	50%	45%	7%	0%	0%	0%	0%
Category for action	Blue	Blue	Green	Yellow	Yellow	Orange	Red

Calibration Rules for Categories for Action on Functionality
These rules may be adjusted, depending on objectives for managing the portfolio.

BLUE = Exceeds need. At least 90% of topics are in acceptable range. **More than 20% of topics** exceed required level of functionality. **All meet required threshold levels.**

GREEN = Good Fit for the functionality requirement profile. **At least 90%** of topics are in acceptable range. **Less than 20% of topics** exceed required level of functionality. **All meet required threshold levels. All exceptionally important topics** are not more than 1 level below requirement.

YELLOW = Consider = At least 70% of topics are in acceptable range. May have some **topics below threshold.**

ORANGE = Deficient = 50% to 70% of topics are in acceptable range. May have some **topics below threshold.**

RED = Dysfunctional = Less than 50% of topics are in acceptable range. May have some **topics below threshold.**

Figure 10.5 How suitable: asset fit calibrated according to action required.

information that can be audited and verified because the scales are calibrated and standard-ized. They do not depend on intuition or professional judgement, but on an assessment of what is physically present on drawings or in the existing facility. These assessments can be completed in a few days by a small team, at costs substantially lower than in-depth facility evaluations, and they provide warnings and pointers for more detailed investigations.

10.3   Applying the methodology to assess the suitability of a portfolio of assets

An organization with a number of property sites wanted to know: 'Do our real estate assets provide the required functionality? Or more? Or less?' The purpose of the project was to assist the organization to allocate resources according to mission-related priorities. Based on the results from a site investigation, the authors and a team of information specialists assembled a visual presentation of the major components of an asset and port-folio analysis in the form of suitability stamps. This provided the organization with the information it needed, summarized in such a way that senior management could see at a glance which assets were mission-critical and whether each asset was capable of supporting that mission. Relevant information about major physical assets at one site was collected and analysed using the ST&M approach, including the ASTM standards described above.

The main components of the suitability stamp, as shown in Figure 10.4, relate the cap-ability of the facilities to the requirements of a major user group at a site with several build-ings, and show how critical the facilities or zones are to its mission. Figure 10.4 summarizes the information available, or estimated, for 10 buildings and/or zones in a building. It shows how these facilities match requirements for five key sets of issues listed below (1, 4, 5, 6, 7) selected from the following:

1. M = *importance to the mission*. Relative importance of each core asset for the mis-sion of the organization.
2. B = *building codes and regulations*. Estimate of fit between the levels required and the actual situation: occupational safety and environmental health; fire and life safety; accessibility; seismic and blast resistance; etc.
3. E = *environmental protection*. Estimate of fit between the levels required, and the actual situation. Environmental regulations, hazardous materials and waste, compli-ance, sustainability, life cycle analysis, etc.
4. S = *security*. Estimate of fit between the levels required, and the actual situation. This would be based on actual assessments, such as threat and risk studies, including data about likelihood of natural hazards, such as floods, hurricanes, earthquakes, and crime and terrorism.
5. C = *condition and service life*. Fit between the condition and anticipated remaining service life required to support the mission, and estimated budget risks.
6. F = *functionality*. Assessment of fit between the functionality required for the mis-sion, and the serviceability of the asset.
7. U = *utilization*. Per cent of space utilized, or over-utilized (over-crowding).
8. R = *real property*. Easements, in grants/out grants, title, metes and bounds, etc.

9. $\$ = costs$. Compared against budgets and benchmarks: first project and construction costs, or any of the following separately or in combination: costs for renovations/ repairs and alterations (R&A); fit-up at move-in, or furniture, fixtures and equipment (FFE); response to IT and technology changes, etc.; other costs such as life cycle costs (LLC).

10. $L = location$. The suitability stamp of each asset is linked to its location on the site, via its GIS code.

Figure 10.5 contains a summary of the functional suitability of the 10 major assets on the site and provides an overview of the gap analysis for $F = $ functionality, as categories for action. Figure 10.5 demonstrates how the assets in this example were ranked from best fit at left to greatest deficiency at right. The functionality and serviceability of each asset were classified in accordance with the ASTM standards. For each facility, a category for follow-up action was derived from comparing the functionality required for the mission with the serviceability of the facilities. Functionality and serviceability profiles, developed by the authors over a two-week period on site during the summer of 2003, were based on interviews with key user representatives and on assessments of each facility.

Each building houses several different functions, so each facility or zone has its own serviceability profile. Figure 10.5 indicates that some facilities, such as the maintenance shops in buildings 3 and 4, more than meet requirements for functionality. At the same time, other facilities in building 3, and some facilities in buildings 1 and 4, are functionally deficient, and one facility in building 3 is dysfunctional. In time, the organization expects that all results, and their underlying data, will become part of a computerized dataset for assets on all sites. Personnel with access rights to the database will have access to this information and have it graphically displayed in a GIS that links each physical asset (buildings and infrastructure) to its geographic location.

10.4   Conclusions

In most organizations, the kinds of data and information described in this chapter are stored in separate 'silos', with many disconnects between the different phases of the life cycle of a facility. Too often, data are captured repeatedly, stored in incompatible formats and are difficult to correlate and keep accurate. The use of computerized databases, virtual building models and web-based software applications and projects are steps towards the creation of a shared information base for the management of real estate assets. Once such shared databases exist, the value of evaluations and benchmarking exercises will increase, because the information will be easier to retrieve when needed. The shared knowledge base will make it easier to close the feedback loop, and relate the facilities delivered (supply) to the requirements (demand) of the stakeholders.

An integrated and shared data and knowledge base is a potential source of major savings in the management of real estate assets. More importantly, it will lead to the reduction in misunderstandings, the increased ability to pinpoint weak links in the information transfer chain, and improvements in products and services because the lessons learned will not be lost. The proliferation of tools and methods such as ST&M will, in time, ensure that BPE becomes a routine part of the management of facilities.

Acknowledgements

This chapter is based, in part, on the paper cited below: Szigeti et al. (2004) and summarizes other writings of Szigeti and Davis.

References

Ang, G. et al. (2002). A Systematic Approach to Define Client Expectation to Total Building Performance During the Pre-Design Stage. In *Proceedings of the CIB Triennial Congress 2001 – Performance in Product and Practice*. Rotterdam, Holland: CIB (International Council for Research and Innovation in Building and Construction).

ASTM (American Society for Testing and Materials) (2000). *ASTM Standards on Whole Building Functionality and Serviceability*, 2nd edition. West Conshohocken, Pennsylvania: ASTM (American Society for Testing and Materials).

Baird, G. et al. (1996). *Building Evaluation Techniques*. McGraw-Hill.

Consumer Reports (2000). *Rating the Raters*, August 31, 2000. Consumers Union of USA, Inc.

Davis, G. et al. (2000a). *Serviceability Tools, Volume 7. Requirement Scales for Service Yards*. ICF (International Centre for Facilities).

Davis, G. et al. (2000b). *Serviceability Tools, Volume 8. Rating Scales for Service Yards*. ICF (International Centre for Facilities).

Davis, G. et al. (2001). *Serviceability Tools, Volume 3 – Portfolio and Asset Management: Scales for Setting Requirements and for Rating the Condition and Forecast of Service Life of a Facility – Repair and Alteration (R&A) Projects*. IFMA (International Facility Management Association).

Davis, G. et al. (1993). *Serviceability Tools Manuals, Volumes 1 & 2*. ICF (International Centre for Facilities).

Gibson, E.J. (1982). Publication 64, CIB Report. In *Working with the Performance Approach in Building*. CIB (International Council for Research and Innovation in Building and Construction).

Hattis, D.B. and Becker, R. (2001). Comparison of the systems approach and the nordic model and their melded application in the development of performance-based building codes and standards. *Journal of Testing and Evaluation, JTEVA*. Vol. 29, No. 4, pp. 413–422.

McGregor, W. and Then, D.S. (1999). *Facilities Management and the Business of Space*. Hodder-Arnold.

Meacham, B., Tubbs, B., Bergeron, D. and Szigeti, F. (2002). Performance System Model – A Framework for Describing the Totality of Building Performance. In *Proceeding of the 4th International Conference on Performance-Based Codes and Fire Safety Design Methods* (FSDM & SFPE), 2002. Bethesda, MD: SFPE (Society of Fire Protection Engineers).

Preiser, W. and Schramm, U. (1997). Building performance evaluation. In *Time-Saver Standards for Architectural Data* (D. Watson, ed.), pp. 233–238. McGraw-Hill.

Prior, J. and Szigeti, F. (2003). Why all the fuss about performance-based building? In *CIB PeBBu Website*. CIB (International Council for Research and Innovation in Building and Construction).

Szigeti, F. and Davis, G. (2001a). Appendix A – Functionality and Serviceability Standards: tools for stating functional requirements and for evaluating facilities. In *Federal Facilities Council, Learning From Our Buildings: A State-of-the-Art Practice Summary of Post-Occupancy Evaluation*. National Academy Press.

Szigeti, F. and Davis, G. (2001b). Matching people and their facilities: using the ASTM/ANSI standards on whole building functionality and serviceability. *In CIB World Building Congress 2001 Proceedings: Performance in Product and Practice*, CIB (International Council for Research and Innovation in Building and Construction).

Szigeti, F. and Davis, G. (2002). User Needs and Quality Assesssment. In *Facility Management Journal (FMJ), January–February, 2002*. Houston, Texas: IFMA (International Facility Management Association).

Szigeti, F. et al. (2003). *Case studies*: assessing quality – the successful response to user requirements. In *Proceedings. Toronto, Canada*: World Work Place 2003, Houston, Texas: IFMA (International Facility Management Association).

Szigeti, F. et al. (2004). Defining Performance Requirements to Assess the Suitability of Constructed Assets in Support of the Mission of the Organization. In *Proceedings. CIB World Congress, Toronto 2004*. CIB (International Council for Research and Innovation in Building and Construction).

11

Assessing the performance of offices of the future

Rotraut Walden

Editorial comment

The building performance evaluation questionnaire is often equated with post-occupancy evaluation (POE) in the conceptual framework for building performance evaluation (BPE) (Preiser and Schramm, 1997). How an evaluation of this kind can best be carried out, and the way in which it can be enhanced through detailed user-needs analysis (UNA) is the subject of this chapter. User needs analysis calculates the gap between the condition of a building at the present time and future demands on it.

Results from the BPE questionnaire are compared to those collected by the Koblenz questionnaire, which focuses on the connection between environmental features and psychological data, such as evaluations of work efficiency, well-being and environmental control. From the outcome, it is possible to anticipate future user needs in office buildings. The process leading from evaluation results to strategic intervention is described. Systematic integration of research questions, research methodology and design guidelines are facilitated through the 'facet approach' (Borg and Shye, 1995). The 'facet approach' provides a structure, in the form of a general statement, from which methods for analysis can be derived.

11.1 Increasing productivity in companies through better office buildings

In spite of the high costs of trained staff, companies tend not to invest in environmental improvements to help people work more effectively. Recent studies of the psychological well-being of employees indicate that the work environment is important. Among the ways of measuring the effects of the workspace on employee psychology are the BPE approach, with an internationally applied survey questionnaire designed to measure building performance, and the Koblenz questionnaire, widely tested in Germany, designed to measure user psychology (BPE questionnaire: Preiser, Rabinowitz and White, 1988; Koblenz questionnaire: Walden, 1999). The purpose of this chapter is to show that by using data collected by these methods, companies can make decisions that will help employees become more productive and thereby improve the economic performance of the organization.

The effects of the workspace environment on worker productivity can be calculated as follows. Eighty per cent of a company's office expenditures are staff costs (Schneider and Gentz, 1997). Of the remaining 20 per cent, 12 per cent are for equipment, communications, technology, office supplies, and furniture. At 8 per cent, building expenditures contribute the smallest share. Sixty per cent of these expenditures are construction costs, 11 per cent maintenance, 19 per cent air-conditioning and heating systems, and 10 per cent is spent on cleaning. Brill et al. calculated nearly the same results in 2001 (Brill et al., 2001). Thus the influence of construction costs on office expenditures is marginal; however, construction costs are often the focus of the decision to invest, or not, in a building project because they accrue over a relatively short time. If construction costs are distributed over the building's lifespan and other occupancy costs included, the actual percentage of construction costs is only 5 per cent. During construction, only 5 per cent of the office building budget is spent, but this 5 per cent considerably influences the effectiveness of the other 95 per cent of office spending. If this 5 per cent is badly spent, it is the employees, and by definition the company itself, that suffers. An increase in construction costs of 10 per cent on a building project affects annual space costs at an escalation of 0.5 per cent. This is a sum that could be spent on optimizing work conditions.

One strategy for indicating how new construction or building renovation can improve users' quality of life is BPE. Companies see their employees as their biggest potential growth factor, yet this is the most costly component of doing business. The majority of firms do not take into account that even a minimal investment in workspace can result in a worker productivity improvement of between 10 and 30 per cent. These estimates were produced by Borg (2000, p. 41) in discussing recommendations based on the results of a job satisfaction survey. Gifford (2002, p. 371) reported that an investment in workspace design can result in a productivity improvement of between 10 and 50 per cent; and Brill et al. (1984) calculated a 17 per cent improvement.

11.2   User needs analysis

Previous studies of office building users have focused on the effects of the work environment on user satisfaction (Clements-Croome, 2000). The concept of user satisfaction is presumed equivalent to employee well-being, but other studies have shown that well-being is not always equated with higher work efficiency. Analysis of how a building's performance can respond to 'soft facts', i.e. psychological aspects, can help companies determine how personnel costs (80 per cent) can be reduced and productivity increased at the same time. There are two significant ways of making this determination. First, aspects of the business itself that define quality for employees, for example, upward mobility, individual responsibility, better pay, a safer workplace, and company philosophy. In 1980, an investigation led by the Institute for Applied Social Research (INFAS) in Germany concluded that the most important aspects of job quality are: money-making opportunities; pleasant environmental conditions; job interest; team-work; and (in tenth place), effective worker participation and a higher stake in the workplace. Secondly, workplace ergonomics, for example, a more 'intelligent' office workspace, the design of workspace, and how office design affects people at work offer a valuable line of inquiry.

Both these two influences on research need to be taken into account by identifying methods that enable strategic interventions for future office buildings to be based on the results (Walden, 2003). The focus of this approach is the connection between the quality of the built environment and its performance. A demonstrable link between design recommendations for improving workspace and economic indicators in a company can be established through empirical research into users' needs. Such studies look at how to decrease rates of absenteeism, fluctuation, maintenance, vandalism and theft, and at the same time increase output in terms of corporate goals. The monetary value of each environmental recommendation can be specific to each country, or even for each company (Walden, 2000).

11.3   Methodology

In this chapter, a case sudy of a user needs analysis designed to generate recommendations for strategic intervention is described. The new building for the Academic Training Center for Germany of the Deutscher Herold Insurance Company in Bonn is centrally located in the vicinity of the train station and resembles a castle from the outside. It is surrounded by heavy traffic. Inside the building, Mediterranean palm trees and lush plants await visitors, and there is a garden on the roof. Within the inner courtyard there is a restaurant. Two-person offices line the long, badly-planned floors giving a cramped feeling. The building was completed in January 1998. It accommodates 614 individuals (mostly offices for two people), as well as conference rooms and training rooms on six floors. The study was designed to survey employees in order to determine the effects of the building on work efficiency, and to identify which building elements are important for future building projects.

A facet approach was used in order to develop a system to judge office quality, and used as a basis for the design of the Koblenz questionnaire. A link to the building performance evaluation was created through the comparison of the Koblenz questionnaire with the BPE questionnaire, which was formulated for user needs analysis in office buildings. The response scales of both questionnaires allowed for a direct comparison of the results.

11.3.1   Office quality evaluation data collection

The elements of the system to judge office quality were developed based on the results of interviews containing exploratory questions such as: 'What do you think is the most important … ?' 'Which characteristics of the building as it stands now should be judged positively … ?' 'Which mistakes should be avoided …?' Work environments were then assessed using six criteria, namely functional, aesthetic, social, ecological, organizational and financial criteria (see Walden, in press). Functional aspects save time and energy; for example, layout, wayfinding and quality of materials. Aesthetic design results in feelings of beauty or newness. Social aspects can cause conflicts that arise from simultaneous use of one setting by multiple parties (for example, concentrated work disrupted by someone using a pneumatic drill in the vicinity), or in opportunities for communication. Ecological aspects mean that the environmental consequences of a building's existence are taken into account throughout, from breaking ground to recycling to health concerns. Organizational aspects comprise the space-time breakdown of resources, provision of information, materials and storage facilities for logistics, and sharing methods for task cycling. Financial aspects cater to the possibility of the entire organization's purchases and sales, as well as their production costs. In addition, experts on intelligent buildings were interviewed, and findings from existing research were augmented by students' ideas (Sundstrom and Sundstrom, 1986; Wineman, 1986; Vischer, 1996).

11.3.2   Facet approach

A mapping sentence that provided a basis for the Koblenz questionnaire was initially developed using the facet approach. The mapping sentence is read as a complete sentence in which individual parts of the sentence can be systematically exchanged and combined, as shown in the example given in the Tool kit (see Appendix). The categories of the facet approach are listed in the mapping sentence, and provided the basis for selecting items for the Koblenz questionnaire. The BPE questionnaire had already been tested in previous studies and was applied independently to a survey of building quality.

11.3.3   Using the BPE questionnaire

The BPE questionnaire has 13 response categories and the Koblenz questionnaire has 187 items. Each were filled out on site (in the Deutscher Herold building) by 32 student 'experts' who had completed a special training programme focused on the assessment of office environments using social research methods. The survey compared present building conditions with those aspects considered most important for future buildings. The BPE questionnaire is listed in the Appendix. For this study, additional rating scales were included, for example, differentiating between sound and noise, instead of using just 'acoustics'. The BPE questionnaire also contains global environmental quality measures (lighting, acoustics, temperature, etc.), and it emphasizes spatial quality, such as the quality of building materials used on floors, walls, and ceilings.

11.3.4   Using the Koblenz questionnaire

The Koblenz questionnaire was developed independently from the BPE questionnaire (Walden, 1999). It complements the BPE questionnaire by supplying psychological 'soft facts' on work efficiency, such as employee well-being, and self-regulation of environmental conditions, such as: view; ventilation; heating; cooling; acoustics/noise; self-initiated changes by users; improvements in aesthetic appearance; vandalism rates, etc. In this study, the Koblenz questionnaire was used in the same way by trained researchers.

   The items of the BPE and the Koblenz questionnaires were evaluated according to 5-point rating scales ranging from $+2$ ☺☺ (very good at the moment or very important in the future) to $+0-$ neutral to -2 ☹☹ (very bad or very unimportant), and additionally 'not applicable'. The rating scales were similarly structured in both surveys in order to enable a comparative user needs analysis (Walden, 1999). The Koblenz questionnaire included 'work efficiency', 'well-being' and perceived 'environmental control' by users. Statistical tests were used to calculate the gap between the actual building condition and future importance rankings. If significant differences exist, this signals a call for action for future improvements to the building. Moreover, in each section there were several opportunities for open-ended responses, such as: is there still something missing that has been forgotten, e.g. facilities or services that are needed but not present at this time?

11.4   Selected results

Results from the Koblenz questionnaire complement the results from the BPE questionnaire in the following ways.

11.4.1   BPE questionnaire results

Results from this survey indicate that the most important future need is access for people with disabilities, especially to the gardens. The aesthetic appeal of conference rooms, offices and corridors was seen as unpleasant. Furthermore, the odour of conference rooms could be improved upon. An explanation for this result could be that new furniture is still giving off fumes. In the category of group offices in particular, many improvements are still necessary, for example, adequacy of space, flexibility of use, security, and sound insulation. As in many North American office buildings, acoustic conditions could be improved in the entire building, especially in the single and group offices. Improvements for the future could be made in stairs/corridors, group and single offices, and conference rooms. The rankings are shown in Figure 11.1.

11.4.2   Koblenz questionnaire results

Comparisons between the existing Deutscher Herold building and its future requirements for 'work efficiency', 'well-being' and 'environmental control' result in the following conclusions:

- In terms of productivity, there are still improvements to be made throughout the building, the façade/site, entrance area (significant gap between 'present needs' and 'importance in future'). But immediate improvements are needed in the restaurant/break area, the conference rooms, single and group offices, and the restrooms.
- The level of well-being was judged to be quite variable, although generally, the experts found work conditions satisfactory. Improvements were recommended for the façade/site, the conference rooms, and both the single and group offices.
- 'Environmental control' was assessed to be the worst aspect. Environmental control is expressed as the ability to regulate ambient environmental conditions through participation and personalization. This is the aspect with most room for improvement in regards to the entire building, the façade/site, and the conference rooms. Significant improvements could also be made in single and group offices, and even restrooms.

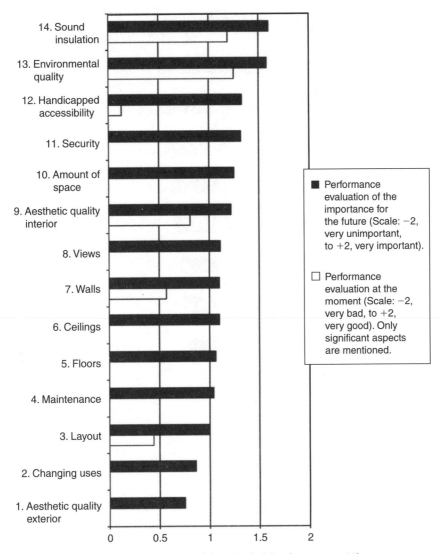

Figure 11.1 BPE questionnaire – overall quality of the entire building (t-tests, p ≤ .01).

Thus, although this innovative Deutscher Herold building has already been judged positively, the survey results indicate the need for some concrete improvements to ensure future quality.

Figure 11.2 shows that the most important environmental elements for the future are the efficiency of the overall building, natural lighting, user well-being and ease of communication. The temperature is already at a satisfactory level, whereas ventilation needs to be improved. Less important for the future, but requiring improvement nonetheless are: access for people with disabilities, occupants' personal modifications for usage, and environmental controls throughout the building.

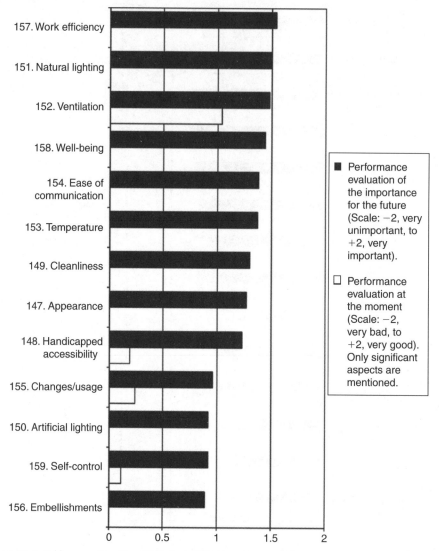

Figure 11.2 Koblenz questionnaire: 10. Entire building (t-tests, p ≤ .01).

11.5 Conclusions

Close examination of results from the two surveys indicates that they are both individually useful and complementary in important ways. They can easily be integrated into one standardized questionnaire in order to compare office building evaluations internationally. Items that could be included in future versions of the BPE questionnaire are:

● ventilation, heating, cooling, sun protection/blind, noise, sound, lighting, operable windows, etc.;

- smoking area;
- storage;
- degree of freedom to control the above items/areas;
- waiting time for elevators;
- climate in winter/summer.

The Koblenz questionnaire has already been revised and tested on the University of Koblenz's facilities. The new version (203 items) includes 17 performance indicators of organizational effectiveness that are affected by the building, as well as salient physical characteristics of the work environment. Collecting data on performance indicators was found to be useful in demonstrating the monetary value of decisions in improving a building to increase productivity.

The facet approach, in combination with user evaluation of office quality, exhaustively and systematically identifies those aspects of future office quality which were acquired through the Koblenz questionnaire, and could also serve to develop future improvements in the BPE questionnaire.

11.6 Summary

The facet approach provided a structure for the walk-through questionnaire and analysis and evaluation of results. On-site observation of facilities by trained experts identified environmental elements necessary in the future, as well as what the building has already achieved. In future studies, experienced users can also be questioned. Results suggest that in Germany, respondents prefer the opportunity to regulate environmental conditions (lighting, heating, cooling, etc.), while at the same time using sensory technology such as fully-automatic central air-conditioning; and they prefer to personalize improvements in aesthetic appearance. In addition, users prefer natural lighting in their workplace, with access to a view to the outside.

The results demonstrate that there exists a high potential to increase worker efficiency by changing the building. Additional analysis shows significant correlations between future work efficiency, well-being in the overall building, and well-being in individual offices. Respondents also equate better control over stress with improved environmental control. Ratings for future environmental control are correlated with more actual control over ventilation, heating, air-conditioning, acoustics, sanitary facilities, and personal improvements made by the occupant to the aesthetic appearance and to the functionality of the entire building.

These results indicate that an increase in well-being leads to an increase in worker productivity, and, according to these results, the reverse is also true, since productive workers often feel better. However, it can also be concluded – in line with the results of previous research – that work efficiency and worker well-being are highly correlated but not equivalent to each other. The results of this study confirm that work satisfaction and work efficiency have different origins: users are already satisfied with the existing situation, but improvements could be made in the building concerning work efficiency; improvements to the building would therefore lead to increased productivity.

Integrating the BPE and Koblenz survey tools provides a comprehensive approach that will be used in future studies to gather data to help architects make design decisions oriented to future as well as present uses. In the long term, companies will be able to provide more appropriate work environments for their employees and thereby improve the economic performance of the organization.

Acknowledgements

Special thanks to the architects from Norman Foster & Partners, Commerzbank, Deutsche Bank, Dieter Wendling & Sons, BIG-Center Birkenfeld; as well as all the students from the University in Koblenz. The author is also grateful to Ruth Rustemeyer and Natalie Fischer from the University of Koblenz for early comments on this draft.

References

Borg, I. (2000). *Fuehrungsinstrument Mitarbeiterbefragung: Theorien, Tools und Praxiserfahrungen* (2. Aufl.). Verlag fuer Angewandte Psychologie. (Instrument of Leadership in Employee Surveys: Theories, Tools and Experiences (2nd Edition). Applied Psychology Press.)

Borg, I. and Shye, S. (1995). *Facet theory. Form and content.* Sage.

Brill, M., Margulis, S., Konar, E. and BOSTI (Buffalo Organization for Social and Technological Innovation) (1984). *Using Office Design to Increase Productivity. Inc. Vol. I and II.* Workplace Design and Productivity Press.

Brill, M., Weidemann, S., Allard, L., Olson, J., Keable, E.B. and BOSTI (Buffalo Organization for Social and Technological Innovation) (2001). *Disproving Widespread Myths about Workplace Design.* Kimball International: Jasper, Indiana.

Clements-Croome, D. (ed.) (2000). *Creating the Productive Workplace.* E&FN Spon.

Gifford, R. (2002). *Environmental Psychology: Principles and Practice* (3rd Edition). Optimal Books.

Preiser, W.F.E. and Schramm, U. (1997). Building Performance Evaluation. In *Time Saver Standards* (D. Watson, M.J. Crosbie, and J.H. Callender, eds) (7th Edition) pp. 233–238. McGraw-Hill.

Preiser, W.F.E., Rabinowitz, H.Z. and White, E.T. (1988). *Post-Occupancy Evaluation.* Van Nostrand Reinhold.

Schneider, R. and Gentz, M. (1997). *Intelligent Office – Zukunftsichere Buerogebaeude durch ganzheitliche Nutzungskonzepte.* Verlag R. Mueller. (Intelligent office – future office buildings through entire usage concepts.)

Sundstrom, E. and Sundstrom, M.G. (1986). *Work Places. The Psychology of the Physical Environment in Offices and Factories.* Cambridge University Press.

Vischer, J.C. (1996). *Workspace Strategies: Environment as a Tool for Work.* Chapman & Hall.

Walden, R. (1999). Work-efficiency and Well-being in offices of the future. Paper presented at the symposium: W.F.E. Preiser, Evaluation of Intelligent Office Buildings in the Cross-Cultural Context. In *Proceedings of the 30th Annual Conference of the Environmental Design Research Association. The Power of Imagination* (T. Mann, ed.) pp. 258–259, Orlando, Florida, June 2–6. EDRA.

Walden, R. (2000). Human resources in office design – or how to estimate monetary outcome from recommendations for future buildings? A model calculation. Paper presented at the Symposium: W.F.E. Preiser, Intelligent Office Buildings. In *Proceedings, Program, Abstracts. People, Places, and Sustainability* (G. Moser, E. Pol, Y. Bernard, M. Bonnes, J.A. Corraliza and M.V. Giuliani, eds). 16th Conference of the International Association of People-Environment Studies. I.A.P.S. Sorbonne in Paris, France, July 4–8.

Walden, R. (2003). Intelligent Offices for the University of the Future. Paper presented at the Symposium: W.F.E. Preiser, Performance Evaluation of Office Buildings. In *Proceedings, Papers, Abstracts. People Shaping Places Shaping People* (J.W. Robinson, K.A. Harder, H.L. Pick and V. Singh, eds) p. 306. Environmental Design Research Association 34, Minneapolis, MN, USA, May 21–25. EDRA.

Walden, R. (in press). Work Environments. In *Encyclopedia of Applied Psychology* (C.D. Spielberger, ed.), Academic Press.

Wineman, J.D. (ed.) (1986). *Behavioral Issues in Office Design.* Van Nostrand Reinhold.

12

Assessing Brazilian workplace performance

Sheila Walbe Ornstein, Cláudia Miranda de Andrade, and
Brenda Chaves Coelho Leite

Editorial comment

In the wake of the transformations that have affected globalized mega-cities, the city of São Paulo, Brazil, has considerably increased its stock of office space, to approximately $8\,500\,000\,m^2$. Today it is considered the largest business centre in Latin America.

Most of the city's office buildings reflect international architecture and stylistic trends, with their glass façades and monumental lobbies, and are considered by developers to be intelligent buildings. In Brazilian cities in general, however, not only in São Paulo, the alleged 'intelligence' of these buildings is limited to their exterior. Inside, the general rule is lack of quality, from occupancy aspects to layout, infrastructure and comfort conditions.

This situation seems to be the result of a lack of foresight and long-term perspective on the part of decision-makers, since the buildings are more oriented toward the companies' corporate image than to their efficiency. There seems to be little or no concern for adaptive reuse/recycling of these buildings. In this context, a POE database is being generated by academic researchers as a starting point for future, systematic building performance evaluations, in order to provide feedback on building quality to all those involved in the process, from building delivery through to the end of the life cycle.

This chapter presents the results of post occupancy evaluation (POE) studies carried out between the early 1990s to the present, focusing on the technical and functional performance of workspace in office buildings. In addition, the chapter discusses how a company's culture is reflected in its physical workplace and space-use, and makes recommendations for building design, construction, delivery and operation throughout its life cycle.

12.1   Background

From 1987 to 1997, the most intense growth of high-rise commercial and service buildings in São Paulo occurred in the south-eastern section of the city, along the Pinheiros River Expressway (*Avenida das Nações Unidas*) (Bruna and Ornstein, 2001). This long parade of tall buildings represents 41.8 per cent of the total commercial land use of the city. Although today São Paulo is a tertiary mega-city, the vacancy rate in its Class A office buildings remains high, approximately $292\,000\,m^2$ in 2002, as compared to $194\,000\,m^2$ in 2001. Nevertheless, there is a lack of office buildings with floor areas of $800\,m^2$ or more, and this severely limits the workspace configuration possibilities for corporate tenants (Ornstein, 1999; Andrade, 2000; Leite, 2003). Moreover, most of the new buildings tend to reflect stylistic 'fashions', with glass façades and monumental main lobbies, while the function and quality of the indoor environment does not meet comparable standards of quality. This is in part due to the characteristics of the Brazilian office real estate industry, since those who build the buildings have no clear commitment to the persons and groups who are going to operate and use them throughout their lifetime. Therefore building exteriors take priority over other important features of the design, which are ignored or eliminated due to deadlines and cost cuts. In addition, corporate decision-makers are themselves more oriented to the importance of their corporate image than to the building's efficiency, and do not consider building performance when buying or leasing facilities.

Since 1995, the São Paolo University research team has tried to remedy this situation by carrying out systematic assessments of office building performance, with an emphasis on workspace quality and feedback from occupants, using physical measurements and technical analyses. The resulting database serves as the basis for a more comprehensive and systematic BPE approach in the future, with an eye to providing feedback on quality to all groups involved in building delivery and operation. This approach to POE was developed as a procedure for assessing building performance for the facility and property management industries (Ornstein et al., 2001). Researchers in other countries have studied a range of issues related to work environments, ranging from the performance of commercial buildings (Duffy et al., 1993) to the impacts of the work environment on productivity (Shumake, 1992; Clements-Croome, 2000). However, the only standard applied to Brazil's commercial buildings is the amount of space per person.

Based on the conceptual framework for building performance evaluation (Preiser and Schramm, 1997), POE procedures were used in all the case studies in order to understand both how companies are physically organized, and to what extent office space can serve to help the organization become more competitive. Two office buildings are the subject of this chapter. Both are considered to be high technology buildings, and were designed and built to be rented to one or several corporate tenants. Today, both house a variety of companies in different market segments, with diverse countries of origin. The results compare how one standardized physical environment responds to different companies' space needs, identifying cultural differences that may or may not be expressed in the different workplace layouts.

12.2   Methodology

The research, based on the building performance evaluation process proposed by Mallory-Hill et al. (see Chapter 15), was carried out in three stages, as shown in Figure 12.1.

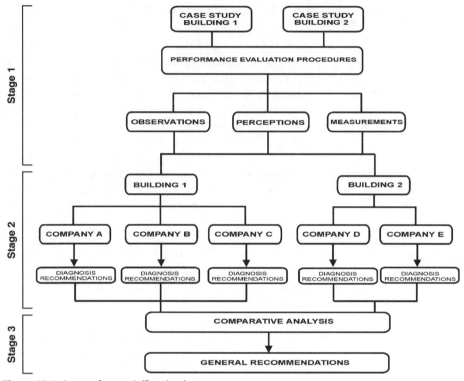

Figure 12.1 Stages of research (flowchart).

The observation phase included an examination of the original plans and the as-built versions, followed by walk-through observation of the specific areas of the building to be studied. Each company was then carefully observed, once at the beginning of the study and a second time on the day the survey questionnaire was distributed to users. In order to gauge occupants' perceptions, unstructured interviews were held with key informants, such as each building's property managers, maintenance and security staff, as well as with the facility manager for each tenant. Information was obtained on the use, operation, maintenance and management of building systems and spaces.

Questionnaires were then distributed to a sample of employees, which varied in number from one corporate tenant to another (see Figures 12.2 and 12.3), making a total of 172 in Building 1 and 85 in Building 2. The questionnaire was first piloted to employees in Corporation B, and then revised. The final questionnaire consisted of items pertaining to respondent characteristics, details about work activities, assessment of building amenities and services, evaluation of the office work environment, furniture and ambient environmental conditions. In addition, open-ended questions invited respondents' comments. On-site measurements of ambient conditions were made on all floors occupied by the target companies, including temperature, relative humidity, lighting and noise. Digital instruments and procedures were used, and results compared to recommended Brazilian and ISO standards (Associação Brasiliena de Normas Técnicas, 1977, 1982, 1990; International Standards Organization, 1985).

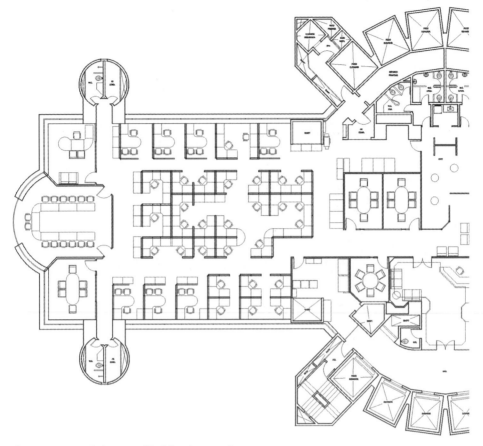

Figure 12.2a Main features of Building 1 companies.

The survey also took into account other key aspects of the work environment, in order to determine the buildings' efficiency. The main space-use categories measured were:

- Individual workspace: total space allocated only to workstations and offices.
- Meeting rooms: workspace where group activities are carried out.
- Support areas: space reserved for resting, leisure or small informal meetings, such as lounges, coffee break areas, and printing and copying pools.
- Main circulation: general distribution of space on floor-plates, including emergency evacuation routes.
- Storage areas: areas for common use, and distributed or centralized to accommodate shared storage.
- Technical areas: for infrastructure and telecommunications.
- Workstation size: dimensions of each individually-occupied permanent workspace.

Based on the results of previous surveys of Brazilian buildings, performance indicators for different types of use were defined. These data, along with user survey responses, were compared to Brazilian norms and regulations, and to the results of other similar studies (Duffy, 1997).

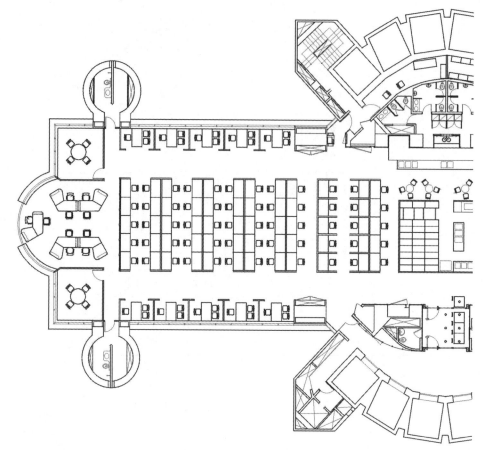

Figure 12.2b (Continued).

12.3   Description of the two buildings

The two office buildings selected were chosen for their common physical features, and because they represent current design trends in Brazilian high-tech buildings.

The two buildings are located in the same neighbourhood (*Avenida das Nações Unidas*), are both 150 meters high and have more than 30 office floors. They are each occupied by about 25 different corporate tenants, mostly large Brazilian or multinational corporations, and are considered high-tech buildings, with Intranet and Internet, automated business systems, automated access control, central air-conditioning and other facilities. The tenant companies were selected for study because of their location; that is, two on low floors (Companies C and D), one on an intermediate floor (Company A), and two on upper floors (Companies B and E). They have different national origins, and are representative of the majority of the companies that occupy the buildings

Figures 12.2 and 12.3 give a general view of the five companies. Their names have been replaced by letters to protect their anonymity.

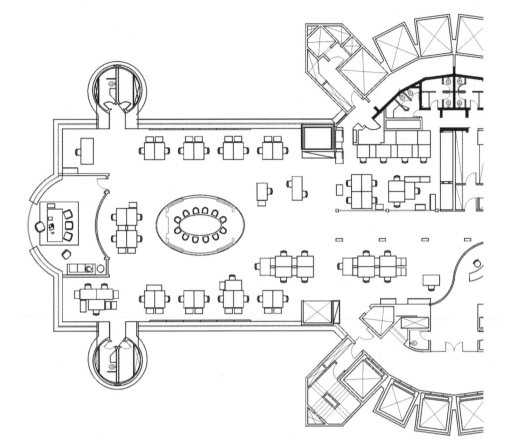

Figure 12.2c (Continued).

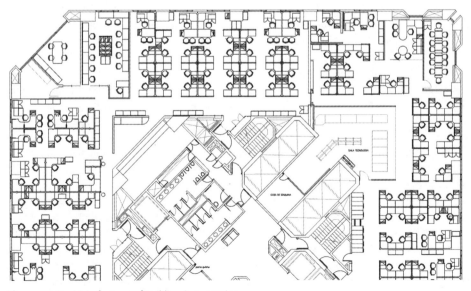

Figure 12.3a Main features of Building 2 companies.

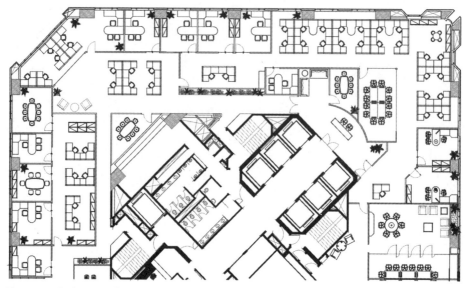

Figure 12.3b (Continued).

12.4 Physical occupancy evaluation

As can be seen in Table 12.1, the five companies differed significantly. Most allocated at least 55 per cent of their floor area (less core space and toilets) to individual workspace, 55 per cent being the minimum standard or performance benchmark in Brazil. As the workstation constitutes the primary work area, percentages below 55 per cent may indicate an inefficient spatial layout or even an inefficient building. As Table 12.1 shows, Company A devotes only 51 per cent to individual work areas. Although this is below the recommended standard, it does not represent a serious deficiency in the workspace layout because 16 per cent is set aside for meeting rooms (compared to the more usual 10 per cent of floorspace). Company E, however, has only 40 per cent of its total area allocated to individual workstations, which is a very poor percentage, even considering the 23 per cent used for meeting rooms. This situation is due to the circulation occupying twice the necessary area, as can be seen in Figure 12.3.

Company C has 21 per cent of its space reserved for primary circulation, far higher than the 12 to 15 per cent standard. Its 3 per cent for storage is far below the norm (8 per cent). The employees here considered lack of storage to be one of their biggest problems. Company E proved to be the most efficient regarding storage space. Although storage occupies only 3 per cent of the building floor area, the survey results indicated a high percentage of satisfaction. Company B has too much space for circulation (26 per cent of the total), to the detriment of other areas, such as meeting rooms and support areas.

Brazilian subsidiaries of American companies showed the highest percentage of support areas (7.6 per cent on average), whereas Company C, which is Brazilian, used 4.5 per cent, and Company B, mixed American and Brazilian ownership, used only 3 per cent for support space. This difference might be the result of different corporate design guidelines for shared workspace. However, foreign standards do not seem to have had a determining influence on space allocation to individual workstations. Typical Brazilian offices are

Table 12.1 Areas dedicated to each use, for each company

Company/ types of use	Performance indicator	Building 1			Building 2	
		Company A	Company B	Company C	Company D	Company E
Circulation	12 to 15%	12.6	26	21	13	20
Work space	>55%	51	55	55	57	40
Storage area	8%	8.1	4.5	3	7	3
Meeting rooms	>10%	16	9	15	10	23
Technical areas		4.3	2.5	1.5	7	5
Support areas	>5%	8	3	4.5	6	9

usually smaller than those found in the United States and Canada, where, according to Vischer (1996), the typical average office space for technical, administrative, and clerical staff is $6.3\,m^2$ (60–70 sq.ft), whereas in Brazil it is only $2.75\,m^2$. In this sample, the smallest workstation size is in Company C (Brazilian), with $1.82\,m^2$ (20 sq. ft), and the largest in Company D (United States), with $4.4\,m^2$ (48 sq. ft).

More variation was seen in office space for middle management. Company B allots the same amount of space to managers as it does to support personnel. In the other companies, middle management workstations varied from $6.10\,m^2$ (66 sq. ft) in Company A (Canadian) to $14\,m^2$ (151 sq. ft) in Company E (United States). Middle management workstations in the Brazilian company (Company C) are completely open or have 1.6 m (5 ft) partitions. Less variation was found in senior management offices, which were an average of $17.70\,m^2$ (197 sq. ft), but are nevertheless well below the $22.5\,m^2$ (250 sq. ft) and up found in the United States and Canada.

12.5   User feedback evaluation

The appearance of the workplace (finishes and colours) was positively evaluated by users in all five companies, with satisfaction ranging from 80 per cent in Company A to 94 per cent in Companies C and E. But their evaluation of spatial comfort was quite different, as shown in Figure 12.4.

This figure shows that Company B's space was the most negatively evaluated, in part due to its layout, which is based on the concept of the 'bullpen' open plan. Small workstations arranged in lines result in high occupational density ($6.15\,m^2$ per person), which is below the minimum of $7\,m^2$ per person required by government regulations (see Figure 12.2) (Ministério de Trabalho, 1990).

Privacy in the work environment is a key indicator of well-being and productivity, and the results of the user survey indicated that privacy was a major issue for these employees, although these findings should be qualified in terms of their different requirements for variable rates of interaction and concentration. For Brazilians, concern for lack of privacy seems to go beyond the need for distraction-free work or the right kind of space for team-work. The high density, the limited workstation size and the open office concept overly expose people to distractions and interruptions that make them feel vulnerable and unimportant. According to Sommer (1973) and Bechtel (1997), each individual needs to create personal space and have command over his or her territory.

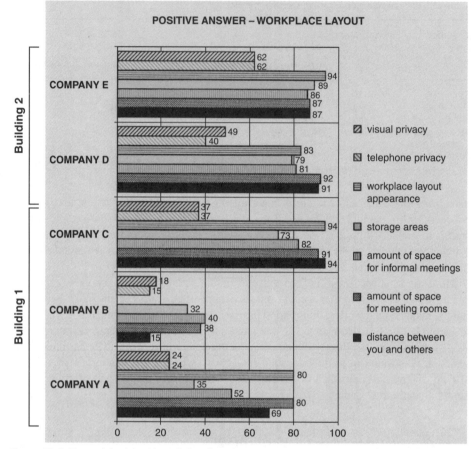

Figure 12.4 User satisfaction with workplace layout.

Privacy was evaluated a little more favourably in Building 2 (see Figure 12.4). Company E indicated more satisfaction with privacy than the others, with 62 per cent satisfied with both telephone privacy and visual privacy. This result is due to a more segmented workspace layout, since the company uses a mix of open and enclosed workspace. Moreover, there is more circulation, creating a lower density (18.6 m²/ws). Company E was also influenced more strongly by its corporate space standards; Company D, in contrast, with the same corporate parent, has a predominantly open layout, and a higher density of 8.1 m²/ws. This may be the reason for 51 per cent of users being dissatisfied with visual privacy and 60 per cent with telephone privacy.

12.6 Environmental comfort evaluation

The evaluation of thermal, visual and acoustic comfort was based on the comparison of data obtained from on-site physical measurements and on the user satisfaction level, as assessed in their responses. Physical measurements were made during working hours on regular work-days.

Measurements of temperature, relative humidity and lighting levels (natural plus artificial lighting) were made at standard locations and under similar climatic conditions. Locations were selected as a function of the geometry of the typical floor-plate and on the concentration of persons and equipment. They were distributed in a relatively symmetrical and repetitive manner throughout the space.

The results are summarized below:

Thermal comfort – The buildings have central air-conditioning systems with ceiling air supply, and constant air volume. The indoor air temperature in Brazilian office buildings is usually kept at 24°C ± 1°C, and relative air humidity (Rh) between 50 and 60 per cent. The results showed that most of the floors in the Building 1 provided thermal comfort and humidity within standard range. In Building 2, the western side of the building is warmer than the other three sides, probably because of direct sunlight on the glass façades. The prevailing air temperatures ranged from 24°C to 26°C, whereas on the southern side the mean air temperature was approximately 3°C lower.

In the two buildings, most of the temperature and humidity readings were found to be within the thermal comfort zone proposed by ASHRAE Standard 55a (1995). However, in Building 1 approximately 30 per cent of the employees complained of feeling 'cool' when the ambient air temperature was around 23°C, and 'warm' when it was over 26°C. In addition, air temperature variations of up to 2°C between areas on the same floor and/or at different times of the day were the subject of complaints. Likewise, in Building 2, 23 per cent of the users said they were dissatisfied with the thermal conditions, feeling 'cold' and 'cool' although temperature readings were 23°C.

Visual comfort – Ratings were positive in both buildings, since more than 80 per cent of the users rated the light level between 'good' and 'excellent'. In Building 1, light levels were relatively homogeneous on all floors, ranging from 400 to 600 lux. In contrast, there were considerable differences in light levels from one area to another in Building 2, with more light at the perimeter and insufficient lighting closer to the core.

Acoustic comfort – Local measurements in two of the companies in Building 1 indicated sound levels consistent with ABNT/NBR-10152 standards, although approximately 40 per cent of the people felt dissatisfied and distracted by ambient noise. In the third company, background sound levels ranged from 45 to 60 dB (A), with peaks of up to 65 db (A) occurring more than 10 per cent of the time. Similarly, 54 per cent of occupants were dissatisfied with noise from voices and equipment. The two companies in Building 2 had an average background sound level of about 45 db (A), with peaks of up to 65 db (A). Complaints from 36 per cent of the occupants in one company (D) and 48 per cent in another (E) referred to intermittent noises. It seems likely that dissatisfaction is more closely related to the lack of privacy resulting from low partitions.

12.7   Conclusions and recommendations

The POE surveys conducted at five different companies in two office buildings provided information on how workplaces perform in relation to different but related aspects of the social, physical, technological and organizational work environment. Recommendations were developed pertaining to improvements in the two buildings' facilities and operations, for example, adjusting air flow on the basis of location of heat sources. This would ensure a more even air flow and more uniform thermal conditions. In addition, lighting controls could improve employees' visual comfort and save energy.

Some recommendations were aimed at stricter compliance with Brazilian standards, such as the issue of circulation areas obstructed by cabinets, or narrower than permitted. The creation of areas for team-work, separated by partitions or high cabinets (1.60 m), was recommended in order to facilitate communication and protect staff from distractions, as well as preventing them from feeling overly exposed. In some cases, suggestions were made to adapt workstations, to reduce problems of lack of storage space and/or loss of concentration, by increasing partition height and providing suspended cabinets. It was suggested that meeting rooms, as well as coffee areas, be placed near main entrances and away from work areas, in order to reduce the impact of noise. The data showed that user complaints about noise do not necessarily result from high levels of background noise, but rather from occasional, disruptive noises.

The level of user dissatisfaction with thermal comfort suggests that the ASHRAE Standard 55a/1995, adopted by the Brazilian ABNT/NBR Standards 6401/1980, is not appropriate for local climatic and cultural conditions. This conclusion is confirmed by recent studies indicating that the maximum air temperature comfort zone for offices is higher than that prescribed by these standards (Leite, 2003). This is an indication that Brazilian standards should be reviewed.

In conclusion, it is strongly recommended that building performance evaluation be applied more systematically to Brazilian office buildings:

a) to create awareness of the forces that impact the building delivery cycle, the building itself, and the workplaces in it;
b) to develop performance standards based on the Brazilian cultural, social and economic reality;
c) to compare with other climates and cultures;
d) to provide information to building decision-makers, thus improving the quality of the buildings, especially inside; and
e) to generate information that will encourage companies to invest in the quality of spaces they occupy. Besides strengthening their corporate image, healthy comfortable office buildings will provide spatial comfort, health and well-being, making employees more productive and, consequently, the company more competitive.

Acknowledgements

In Building 2, Dr Marcelo de Andrade Roméro, Dr Joana Carla Gonçalves, Rafael Silva Brandão and Anna Cristina Miana from the School of Architecture and Urbanism at the University of São Paulo participated in the measurements and analysis of environmental comfort and energy use. Thanks to SATURNO Planejamento, Arquitetura e Consultoria, who provided data on 509 000 m^2 (rentable) in 114 different office buildings in various states in Brazil.

References

American Society of Heating, Refrigerating, and Air-Conditioning Engineers, Inc. *ANSI/ASHRAE Standard 55a–1995, Thermal Environmental Conditions for Human Occupancy.* Atlanta, USA.

Andrade, C.M. de (2000). *Avaliação da ocupação física em edifícios de escritórios utilizando métodos quali-quantitativos: o caso da Editora Abril em São Paulo*, unpublished Master's thesis, Faculdade de Arquitetura e Urbanismo da Universidade de São Paulo, Brazil.

Associação Brasileira de Normas Técnicas. Normas Brasileiras. *ABNT/NBR-5382/1977 – Verificação do Nível de Iluminamento de Interiores – Método de Ensaio*. Rio de Janeiro, Brazil.

Associação Brasileira de Normas Técnicas. Normas Brasileiras. *ABNT/NBR-6401/1980 – Instalações Centrais de Ar-Condicionado para Conforto – Parâmetros Básicos de Projeto*. Rio de Janeiro, Brazil.

Associação Brasileira de Normas Técnicas. Normas Brasileiras. *ABNT/NBR-5413/1982 – Iluminância de Interiores*. Rio de Janeiro, Brazil.

Associação Brasileira de Normas Técnicas. Normas Brasileiras. *ABNT/NBR-10152/1985 – Níveis de Ruído Aceitáveis*, Rio de Janeiro, Brazil.

Bechtel, R.B. (1997). *Environment and Behavior. An introduction*. Thousand Oaks, California: Sage Publications.

Bruna, G.C. and Ornstein, S.W. (2001). *Urban Form and the Transformation of Office Buildings: the São Paulo metropolis case study, Brazil*. Anais do International Seminar on Urban Form, College of Design, Art, Architecture and Planning, University of Cincinnati, Ohio, USA, pp. 150–152.

Clements-Croome, D. (2000). *Creating the Productive Workplace*. London, UK: FN Spon.

Duffy, F. (1997). *The New Office*. London, UK: Conran Octopus Limited.

Duffy, F., Laing, A. and Crisp, V. (1993). *The Responsible Workplace*. London, England. Butterworth Architecture.

International Standards Organization – ISO 7726/1985 – *Thermal Environments – Determination of the PMV for measuring physical quantities*. Geneva, Switzerland.

International Standards Organization – ISO 7730/1994 – *Moderate Thermal Environments – Determination of the PMV and PPD indices and specification of the conditions for thermal comfort*. Geneva, Switzerland.

Leite, B.C.C. (2003). *Sistema de Ar condicionado com insuflamento pelo piso em ambientes de escritórios: avaliação do conforto térmico e condições de operação*. Unpublished Doctoral dissertation, Escola Politécnica da Universidade de São Paulo, Brazil.

Ministério do Trabalho. *Norma Regulamentadora – NR17 – Ergonomia (Portaria 3751 de 23/11/1990)*, Brasília, DF, Brazil.

Ornstein, S.W. (1999). A Post-occupancy evaluation of workplaces in São Paulo, Brazil. *Environment and Behavior* (Vol. 31, No. 4), pp. 435–462. Thousand Oaks, C.A., USA: Sage Publications.

Ornstein, S.W., Leite, B.C.C. and Andrade, C.M. de (1999). Office spaces in São Paulo: post-occupancy evaluation of a high technology building. *Facilities* (Vol. 17, No. 11), pp. 410–422. West Yorkshire, England: MCB University Press.

Preiser, W. and Schramm, U. (1997). Building Performance Evaluation. In *Time-Saver Standards for Architectural Design Data. The Reference of Architectural Fundamentals*. USA: McGraw-Hill, 7th edition, pp. 233–238.

Shumake, M.G. (1992). *Increasing Productivity and Profit in the Workplace*. New York: John Wiley & Sons, Inc.

Sommer, R. (1973). *Espaço Pessoal*. São Paulo: Editora Pedagógica e Universitária Ltda/Editora da Universidade de São Paulo.

Vischer, J.C. (1996). *Workplace Strategies: Environment as a tool for work*. New York: Chapman & Hall.

13

User satisfaction surveys in Israel

Ahuva Windsor

Editorial comment

Office buildings provide the physical context for organizational processes, such as professional activities of individuals and teams, information sharing, and formal and informal meetings. Places inside the building acquire meaning as individuals and groups utilize them and develop habits and a history around them. With time, these places become connecting channels between individual workers and their organizational unit (e.g. large conference rooms are where strategic business policies are discussed, small meeting rooms are where decisions are made, smoking zones are the places for social exchanges, etc.). This process, termed placemaking (Shibley and Schneekloth, 1996), produces meanings that integrate into the overall building knowledge that is carried by its users and sought through user satisfaction surveys. This chapter describes a post-occupancy evaluation (Phase 5, occupancy, feedback loop in the BPE framework) in a specific social and cultural context. The results show how user attitudes towards placemaking affect people's perception and assessment of the built environment.

13.1 The organizational context of post-occupancy evaluations

This chapter describes a post-occupancy evaluation (POE) that evolved out of recognition of the connection between individual and organizational experiences, and knowledge of place, in a new office building. POEs are typically used to assess systematically whether occupied buildings actually work for their operators and users (Preiser, Rabinowitz and White, 1988), and are an important feedback loop in the BPE framework (see Chapter 7). The information obtained from POE is useful to improve future design, construction and facility management. Moreover, users' evaluations of a building reflect its meaning as place, as well as being based on its functional utilization. Such evaluations are done through user satisfaction surveys, one technique used in post-occupancy evaluations and described by Preiser and Schramm (1997).

Two basic assumptions are at the basis of the present work:

a) Post-occupancy evaluation should employ an organizational perspective. This perspective is reflected in contents and methodology, as well as choice of participants and stakeholders. For example, alongside familiar topics such as users' assessment of physical elements and building systems, additional topics for potential study are the social-organizational implications of the design strategy and utilization patterns of group places.

b) Building users in workplaces are primarily workers, and as such, respond to user satisfaction surveys as individuals and as members of organizations. Their responses to questions about building use are affected by individual, group and organizational experiences in the building. They convey meanings derived from their experiences with colleagues in various places in the building.

Any physical change in workspace (building renovations, or moving into new space) is related to organizational change, even if one is not formally declared (Vischer, 1996). For example, the administrative team which gradually replaces personal and departmental secretaries is part of the current trend towards flatter hierarchies and increased team-work. Often, this change occurs as part of new workplace design and job design, in which team and collaborative work plays a more important role. Consequently, individual responses to user satisfaction surveys incorporate both physical and organizational perspectives and issues. These perspectives are inseparable, much like the workplace environment itself which constitutes a system of inseparable elements (Becker and Steele, 1995).

User satisfaction surveys are often conducted before or after the introduction of new workspace. When a new workplace is implemented without adequate preparation, workers are usually preoccupied with new ways of performing tasks and are concerned about the potential cost to their quality of life, and even their employment. Typical questions asked include: 'Is management trying to save money at our expense?' and 'Are layoffs expected as part of the physical change?' Following occupancy, overt or subtle comparisons of the new place to the old are likely to be voiced, often with irritation at an unfamiliar reality and the need to adjust to it. All of these constitute the organizational context of the user satisfaction survey and need to be recognized as such; ignoring it may decrease its contextual validity and weaken the conclusions drawn from the results (Winkel, 1991).

13.2   The government centres project

In 1990, a decision was made at the Ministry of the Treasury in Israel to build nine new government centres in order to increase efficiency in government offices and improve service to the public. By then, most government offices resided in old buildings, many at makeshift locations, with outdated building systems. Most buildings were unsuitable for modernized information technology, and could not support project-oriented work. They were designed for individual clerical work, and their shortcomings interfered with efficient service to the public.

The desired change in the way government employees work contained an element of organizational change, which project leaders perceived as difficult to introduce directly, for structural reasons. Thus, it was decided to introduce improved work environments that would gradually lead to the desired changes in work methods and organizational culture, while improving quality of working life for government employees.

13.2.1 The design strategy

A design strategy for the new government centres was therefore established to reflect the desired changes in ways of working, while preserving certain specified elements of organizational culture. The main objectives of this strategy were increased efficiency, communication, and transparency. The new design was to help facilitate contact between individuals, teams and different government offices, through proximity, openness, and shared facilities. The buildings that have been built are high quality, technologically advanced, and aesthetically pleasing. Floor sizes of the nine buildings range between 12 000 and 30 000 sq. ft, averaging about 25 000 sq. ft (2 300 m²).

Practical implications of the design strategy were:

– allocation of open workstations (70 per cent) and enclosed offices (30 per cent) determined by rank, as is office size;
– all workers have their own desks (as individual territories); workstations are divided by 130–170 cm (65-inch) high partitions; no desk sharing was planned;
– professional services and recreational facilities to be shared by all workers, regardless of rank or ministerial affiliation.

A related strategic decision concerned workers' participation in building design. The sheer size and complexity of the project, and a history of turbulent labour relations between ministry directors and labour unions, led to a flexible implementation approach that left the practicalities of participation in the hands of each project director. Consequently, the extent and nature of participation varied greatly among stages in the design process, topics, and ministries.

Throughout the project, workers' representatives vehemently objected to the design strategy of an open space office layout. They made continuous efforts to expand the allocation criteria of enclosed offices. That way, in addition to rank, functional aspects of job performance requirements would be a basis for allocation of enclosed offices. While in several cases these attempts were successful, the main pillars of the design strategy remained unchanged, and about 60 per cent of workers were allocated open workstations as planned.

13.2.2 The government information agency

The user satisfaction survey described in this chapter was administered to staff of the government information agency (GIA), which moved to its new home as part of its transformation into a public agency. In offering professional services to both public and private clients, GIA had to increase the efficiency and quality of its services in order to compete with similar organizations. Of all the buildings in the government centres project, this was the first to be occupied. As a single occupier building, the organization had greater control over its design and management than the other, multi-occupant centres.

The 550 GIA workers underwent a 12-month process of orientation and preparation that was funded jointly by the treasury ministry and the GIA. A central goal of the orientation phase was to ease workers' adjustment to the open plan office and its social and functional implications, such as reduced privacy, and more interruptions. The process ended with a work-like simulation activity and group discussions that yielded a 'code of conduct' for the open space office. This document was distributed to all workers at move-in.

A last minute strike was threatened by the local chapter of the Workers' Labour Union just before moving in. It was averted as a result of management's commitment to local chapter representatives to conduct a user satisfaction survey within the first year, to assess workers' well-being and how well they functioned in the new workspace.

13.3   Post-occupancy activities

Post-occupancy evaluation, a central component of investigative efforts to assess building performance following occupancy, is usually conducted during the first 12 months of occupancy. It often combines several techniques that are utilized simultaneously and complement one another in order to create a comprehensive picture of the effectiveness of the building. At the GIA, the following post-occupancy evaluation activities were carried out during the first 12 months in the building:

1. A 'Complaints committee' comprising management and workers (including maintenance unit) representatives, received requests and complaints concerning the building. Open space-related interruptions and indoor air complaints constituted about 30 per cent of calls.

    In response, moveable partitions were added across openings into all workstations to allow their closure as needed by owners.
2. Independent of requests and complaints, attempts continued to test and adjust the HVAC to increase workers' comfort, with only partial success.

By the end of the first year, the user satisfaction survey was initiated by management in order to fulfil its commitment.

13.4   The user satisfaction survey at the GIA

13.4.1   Context for initiation of the user satisfaction survey

Customary objectives for seeking feedback about a building are improvement of future construction and of present working conditions. Both evolve out of 'organizational learning' (Zimring and Rosenheck, 2001), a concept describing organizations' efforts to improve operations based on evaluation and feedback. Feedback is affected by organizational culture, that is, the combination of assumptions, values, norms, beliefs and behaviours accepted by all members of the organization (Schein, 2000).

In the case of the GIA, and all government and public sector agencies in Israel, seeking information about working conditions in a building places the user satisfaction survey in the context of labour relations and negotiations. At the GIA, working conditions in the new building, particularly the open space office layout and the sealed windows, had been issues of dispute throughout the design and construction phases.

When the user satisfaction survey was initiated, its objectives were formulated by management and workers as follows:

- inquire about workers' well-being in the building.
- identify building elements that can be improved, after a period of regular use.
- open an on-going communication channel between workers and management.
- evaluate the design strategy and its implications on users and on work performance.

13.4.2   Contents of the user satisfaction survey at the GIA

Workplaces are systemic settings, where physical, technological and social-organizational factors are intertwined and affect each other (Becker and Steele, 1995). The contents of the user satisfaction survey questionnaire sought to unravel workers' individual and collective views of the physical, technological and spatial features of the building. Hence, the survey questionnaire was made up of three sections:

- individual workspace: questions focused on individual workers' satisfaction ratings of features in their own area, such as furniture design, storage, lighting, noises and general feeling at the workspace.
- building: questions targeted evaluations of the building as a whole, and various separate components and systems, including security, smoking zones and arrangements, and meeting rooms.
- social-organizational scene: questions examined impacts of the building and its design and management on the organization and on worker behaviour.

This approach acknowledges that workers experience the building both as individual users and as members of organizational groups. In addition, open-ended questions were included to enable respondents to volunteer comments on the open-plan workspace.

13.5   Findings of the user satisfaction survey at the GIA

The survey results offer several examples that connect building design and systems, and the organizational culture.

13.5.1   The design strategy: open workstations

Respondents who were allocated open workstations were divided equally between those reporting a 'good feeling about my workstation most of the time' and a 'negative feeling most of the time', while the vast majority (89 per cent) of those in enclosed offices reported a good feeling most of the time. Positive feelings were related to aesthetics and design functionality; negative feelings with loss of privacy, the HVAC system, and noise levels. Satisfaction with enclosed offices was linked to control over the workplace. Typical comments included: 'I can always close the door if it's noisy'; 'It is possible to turn the light on/off, or have private telephone conversations.'

13.5.2   Occasional informal meetings with colleagues

Informal meetings in public areas between colleagues characterize the GIA culture, and can be seen as evidence for its informal atmosphere, according to workers' statements in orientation group discussions. The new design strategy aimed to preserve this social mechanism for business and professional purposes. The user satisfaction survey thus included questions about employees' perceptions of these meetings and their frequency.

Findings indicated that frequency of occasional meetings within units increased in the new building, and they were also rated more important than meetings between units. Frequency of the latter decreased in the new building. Thus, from a social-organizational perspective, the interior design layout is satisfactory to workers, i.e. the more important meetings – those with same-unit colleagues – take place more frequently in the new building than in the old location.

Informal meetings with colleagues can also be disruptive, particularly as they often take place in close proximity to open workstations. The investigation of this behaviour aimed at uncovering its dual meaning, both as a valued mechanism to help preserve informality in inter-organizational communications, and as disruptive behaviour to those in open workstations.

13.5.3   Disruptive behaviours

To assess the perceived disruption caused by normative behaviours, both informal meetings with colleagues and peeking into open workstations, which were specifically addressed by the Code of Conduct, were examined in the user satisfaction survey.

Peeking into others' workstations was rated as a frequent, and highly disruptive behaviour. In their open-ended responses, employees related this disruption to lack of privacy. Reduced privacy was stated as a prime reason for dissatisfaction with individual workspaces in this study. Although occasional meetings near workstations are frequent, these are not perceived to be as disruptive as peeking into workstations; respondents were almost equally divided between those interrupted by occasional meetings nearby, and those who were not.

In conclusion, these results indicate that the physical environment at work supports a given organizational culture by facilitating or interrupting normative behaviours. Therefore, when the physical environment is modified, the need for employees to adjust to the new environment means a change in organizational culture through adoption of new norms, as well as a modification of the relevant behaviours. At the GIA, this means an end to peeking into others' workstations, and replacing it with a new, normative behaviour.

The user satisfaction survey helped identify discrepancies, or lack of fit, between elements of the physical environment of the office and components of organizational culture, such as norms and behaviours. The user satisfaction survey can also indicate the extent to which the discrepancy itself is disruptive to members of the organization.

13.5.4   Building security system

Computerized building systems offer information about building users that was previously unavailable, but carry with them a risk for misuse in ways that have organizational repercussions. Technologically-advanced building security systems offer information about employees' attendance, their whereabouts in the building, length of time spent at each location and, sometimes, the nature of their activities. Constant use of magnetic cards and biometric devices, as well as cameras, contributes to a reduced sense of privacy. Employees, as building users, see the potential impact of information from these devices as adversarial, as they may lead to salary deductions for lateness and supervisory control over social encounters. The potential for exploitation of information from security systems is all too evident to employees who identify power relations in organizations as being embodied in this ostensibly benign building system.

At the GIA, the advanced security system became almost immediately an issue between management and workers' representatives, the local chapter of the labour union. Following results from the user satisfaction survey, which indicated the extent of workers' concerns for privacy and misuse of data, two actions were taken, although these only offered a partial solution. First, the old system of separate attendance and security cards was reintroduced, and second, a separate, non-computerized reporting system for building exits during working hours was created.

In summary, the reported increase in the sense of safety in the building from the advanced security system was apparently compromised by concerns for increased control over employees' actions.

13.5.5   Smoking arrangements and zones

Smoking arrangements in Israeli workplaces refer to space management, i.e. areas designated as 'smoking zones', equipped with enclosures and ventilation measures, as well as to behavioural requirements. Moving from enclosed offices into open workstations implies changes in office norms and individual behaviour related to smoking. Since almost 30 per cent of workers reported that they smoke during working hours, an acceptable solution was sought to satisfy both smokers and non-smokers, within the dictates of the national law against smoking in workplaces.

The proposed solution was to designate one of the three kitchenettes on each floor as a smoking zone, and equip it accordingly, with a heavy door and a heavy-duty ventilation system. Yet within a few months new arrangements had to be made because of open office occupants' complaints of smoke in their workspace. So the emergency stairs were furnished at each floor level with benches and a smoke exhaust.

Dissatisfaction with the new smoking arrangements was revealed in the rating scales of the user satisfaction survey findings, as well as in the open-ended responses. 'Humiliation of the smokers' was the reason given by many non-smokers as the principal cause of their dissatisfaction. Respondents also expressed concern about smokers' tendency to leave the doors between the floor and the smoking zones open, thereby letting the smoke into the open-office areas. Subsequently, a pleasantly furnished outdoor roof-top sitting area replaced the seven enclosed smoking zones. Apparently, not separating out the smokers from the non-smokers proved more satisfactory than proximity of the smoking zone to work area. With time, the new smoking area became a truly supportive amenity for informal meetings, in line with the original intention of the new design strategy.

The process the GIA underwent in seeking a suitable smoking area solution indicates that satisfaction with spaces in the building is rooted not only in individual preferences, but also in group norms and behaviour. The new open space on the rooftop has an organizational role and meaning: it satisfies smokers and non-smokers alike, and fits into the cultural norms of informality and removing barriers to informal communication.

13.6   Conclusions

Overall, survey responses indicate that both individual workspace and the building systems are individually and organizationally meaningful. Both levels of meaning need to be taken into account when user satisfaction surveys are conducted, to avoid problems arising from fragmented findings that ignore the organizational norms, habits and activities embedded in individual workers' responses.

In the GIA example described here, users' assessments of places and building elements go beyond the individual worker's concerns. The quest for a suitable smoking arrangement focused on a solution that would satisfy smokers as well as non-smokers; the struggles over the security system represented a general threat of invasion of privacy by technology. These, and other findings, indicate that the building affects the organization as a social unit, as well as affecting relations among its members and the behaviour of individuals. Post-occupancy evaluation needs to examine building use at all these levels, ensuring that an organizational perspective is incorporated into all stages of building performance evaluation. Evaluation of design strategy, as well as the resulting spatial layout and management issues, are important feedback loops in BPE.

This study also demonstrates the value of incorporating culture-specific elements into building evaluation. Where labour is organized, the union is party to the initiation or conduct of the user satisfaction survey. In the Israeli government centres project, the union led resistance, and later, conciliation, thus enabling orderly occupancy. In all likelihood, this involvement affected workers' collaboration with, and responses to, the user satisfaction survey.

Finally, an organizational perspective on user satisfaction surveys has implications for human resources departments: workers' well-being is affected by physical conditions in the workplace, which in turn are controlled by facility managers. Human resources departments need to expand their perspective to include this realm as well, and to share with facilities managers such useful tools as the user satisfaction survey to periodically assess workers' well-being in the physical environment of the workplace.

References

Becker, F. and Steele, F. (1995). *Workplace by Design: Mapping the high performance workscape.* Jossey-Bass.

Preiser, W.F.E. and Schramm, U. (1997). Building performance evaluation. In *Time Saver Standards: Architectural Design Data* (D. Watson, M.J. Crosbie and M.J. Callendar, eds), pp. 232–238. McGraw-Hill.

Preiser, W.F.E., Rabinowitz, H.Z. and White, E.T. (1988). *Post-Occupancy Evaluation.* Van Nostrand Reinhold.

Schein, E. (2000). Organizational culture. In *Organization Development and Transformation* (W. French, C. Bell and R. Zawacki, eds), pp. 127–141. McGraw-Hill.

Shibley, R.G. and Schneekloth, L.H. (1996). Evaluation as placemaking: Motivations, methods, and knowledges. In *Building Evaluation Techniques* (G. Baird, J. Gray, N. Isaacs, D. Kernohan and G. McIndoe, eds), pp. 15–23. McGraw-Hill.

Vischer, J.C. (1996). *Workspace Strategies: Environment as a tool for work.* Chapman & Hall.

Winkel, G. (1991). Implications of environmental context for validity assessments. In *Handbook of Environmental Psychology* (D. Stokols and I. Altman, eds). Krieger Publishing Company.

Zimring, C. and Rosenheck, T. (2001). Post-occupancy evaluations and organizational learning. In *Learning from our Buildings: A state of the practice summary of post-occupancy evaluation.* Federal Facilities Council Technical Report No. 145, pp. 42–53. National Academy Press.

14

Building performance evaluation in Japan

Akikazu Kato, Pieter C. Le Roux, and Kazuhisa Tsunekawa

Editorial comment

This chapter presents the results of investigative-level Post-occupancy evaluation (POE) case studies in a recently completed office tower in Nagoya City, Japan. This study of Japanese workplace environments forms part of an ongoing project of the international building performance evaluation (IBPE) research group. This case study of building performance evaluation (BPE) focuses on a culturally diverse context; it illustrates Phase 5 of the conceptual BPE framework – occupancy (Preiser and Schramm, 1997).

The workplace environments discussed here present a unique approach to employing innovative workplace planning and design methodologies in order to address problems related to employee and work style diversity. Employees were surveyed on their functional and behavioural responses to alternative workplace design strategies, such as non-territorial workplaces (see also Chapter 15). Study results illustrate how innovation, in terms of the perception and allocation of workplace environmental resources, can potentially increase the level of functional efficiency of the overall workplace environment. The research methodology was designed to collect data on employees' work characteristics, movement and circulation, and communication behaviours, as well as their evaluation of the workplace environment according to BPE indices based on Preiser's classification of nine performance criteria (Preiser, 2001).

In addition to these objectives, similarities in the physical (built) environments of the selected workplaces provided an opportunity to investigate the effectiveness and applicability of diverse approaches to workplace 'placemaking' (see also Chapter 13). Results were compared to determine:

1. differences in workplace environmental placemaking methodology;
2. differences regarding subsequent workplace and workspace utilization behaviour;
3. the median building performance level for workplaces in the selected building environment.

The data contributed to existing databases and benchmarked indices of building (workplace) performance within the Japanese context.

14.1   Introduction

The emergence of alternative work styles and shared workspace for highly mobile employees necessitates a re-evaluation of space standards in the workplace. Various viewpoints on workplace planning and design have categorized the planning and design of the workplace in terms of layout typology, workspace allocation (private, shared, free-address), and employee work style diversity. Accordingly, the goal of the modern office is not to homogenize the office environment, but to allow for functional diversity in work practices and processes. The quality of any workplace environment, therefore, lies in its ability to provide employees with access to the resources they need to do their work effectively. This clearly implies the improvement of workspace utilization in order to accommodate the increasing number and complexity of user requirements. This has some important implications for the organization in terms of how and where people work.

Growth and change in modern officing concepts have culminated in the design of spaces characterized by fluidity of function and spatial mobility, beyond the traditional physical parameters of the organization. Research in Japanese workplace environments indicates that employee mobility and communication in the workplace are highly related – often occurring in the form of chance meetings between employees moving to and from their respective workspaces. Based upon this research, an alternative approach to workplace placemaking, based on the multiplicity of modes of communication. This conceptual shift in criteria for evaluating workplace environmental performance requires a re-evaluation of the means available to the researcher for adequate collection of data.

In this chapter, the authors explain how BPE methodology in Japan has evolved to address these issues. Workplace mapping – an advanced means of obtaining information on employees' workplace behaviour through observation – is explained, followed by a case study of an innovative Japanese workplace where workplace mapping was used to evaluate building performance.

14.2   The evolution of workplace quality standards

Performance evaluation in Japan, specifically with regards to workplace environments, has received a good deal of attention since the inception of NOPA (New Office Promotion Association of Japan) and the NOPA New Office Award in June 1987, followed by JFMA awards (Japan Facility Management Association) in November of the same year.

The business objectives of NOPA include conducting studies and researching topics related to the comfort and functioning of the office. The NOPA minimum standard was introduced in May 1995 as a measure to improve not only workplace environmental quality, but also workplace productivity. The space standard addresses the general working area and, where necessary, ancillary spaces are also included. In addition to the physical aspects of the workplace, the standard also covers operational and organizational aspects of the workplace environment (JFMA, 2003).

The NOPA minimum standard consists of 22 standards, which are divided into 10 categories, ranging from the general attributes of the workplace, to measures for the provision and use of resources such as office automation (OA) equipment, desks and chairs, and criteria for facilities management, social welfare, and environmental awareness. Target performance levels for all of these standards are set in such a way that they can be applied and interpreted without the need for special training. The NOPA minimum standard can be applied to Phase 5 (occupancy) of the BPE framework, and serves as an important post-occupancy evaluation tool in determining workplace environmental quality within the Japanese context. By categorizing the 22 standards of the NOPA minimum standard in terms of BPE performance criteria and categories, it is possible to illustrate how the NOPA minimum standard can be used in building performance evaluation (see Table 14.1).

The standard and its application have, however, several weak points. Firstly, since the new office minimum is a general standard, all other standards pertaining to specific working conditions within the workplace environment have to be treated separately. Secondly, since it is a general standard, organizations have no legal obligation to adhere to the minimum stipulations. Organizations employing these standards do so out of their own commitment to the improvement of workplace environmental quality, as well as to the improvement of the health, comfort, and well-being of employees. A third weakness of the standard relates

Table 14.1 Comparison of BPE performance levels and the NOPA new office minimum standards

Levels of client goals and user needs	BPE performance levels	NOPA new office minimum
Technical	Health	Standard 5: Break areas (rest areas) Standard 6: Lighting environment Standard 7: Glare prevention measures Standard 9: Air quality Standard 13: Occupant health
	Safety	Standard 11: Safety considerations Standard 14: Electric wiring layout
	Security	
Functional	Functional	Standard 1: Adequate workspace area Standard 2: Width of main circulation Standard 4: Workplace layout Standard 10: Management of operation Standard 15: Desks Standard 16: Chairs Standard 17: Storage furniture
	Efficiency	Standard 18: Facility management Standard 21: Energy conservation Standard 22: Filing and referencing procedures
	Work flow	
Behavioural	Physiological	Standard 3: Ceiling height Standard 8: Background noise Standard 12: Space comfort
	Social	Standard 19: Female employees Standard 20: Aged, people with disabilities
	Cultural	

to the formulation of the 22 standards on the basis of a suggested minimum. This may result in organizations interpreting them as accepted industry standardizations, thereby giving a false indication of the exact level of workplace environmental quality in Japan.

14.3   BPE methodology in Japan

In Japan, the concept of building performance evaluation (BPE) has become a popular means of determining the qualitative and quantitative aspects of person-environment relationships. By employing the principles of BPE, in combination with various techniques of building performance evaluation, evaluation measures can be improved, and workplace environments assessed, benchmarked, and compared both locally and on a global level.

BPE in Japan is characterized by an extensive investigation of employee activity and communication within the workplace environment. Research in Japanese workplaces has found diversity of workplace communication, in terms of the frequency, extent, and modes, to be a prominent characteristic of the increased focus on knowledge work in modern workplace environments. BPE methodology in Japan has evolved into a way of evaluating environmental design on the basis of the various modes of communication observed, rather than on the basis of individual space standards. This approach does not necessarily discount the importance of space standards in workplace planning and design. Instead, given the high level of employee mobility in the workplace, it has become clear that individual space standards on their own are not sufficient to explain workplace environmental characteristics. Thus, BPE and post-occupancy evaluation (POE) in Japan has focused on diversity of employee communication behaviours as a means of investigating, documenting, and benchmarking relationships between users and their workplace environment. Data obtained through BPE/POE provide workplace planning professionals with more objective information on employee diversity, and facilitates the functional programming, planning, and design of workplace environments that are more responsive to modern concepts of work.

Means of collecting data include questionnaires which are distributed to all employees in the workplace, focus groups and interviews with selected employees, video footage of the workplace environment (where possible), and a workplace survey using various mapping techniques. Through various investigations in Japanese workplaces, the specific methodology applied to the mapping of the workplace has evolved into an intensive documentation, or mapping, of all employees' behaviour in terms of the following:

- Activity mapping – the nature of each individual employee's activities in the workplace.
- Activity duration mapping – the amount of time spent on various workplace activities, as well as the amount of time spent both in the workplace and at individual workspaces.
- Movement mapping – the extent, frequency, and characteristics of employees' movement in the workplace environment.
- Communication mapping – the extent, frequency, and characteristics of employees' communication behaviour in the workplace.

The use of mapping methodology during surveys in workplaces has been very successful. Data collection using mapping procedures, in conjunction with questionnaires and interviews, has yielded advantages over other methods. Mapping of the workplace is done on a mapping sheet of the spaces to be surveyed. The mapping sheet contains a layout of

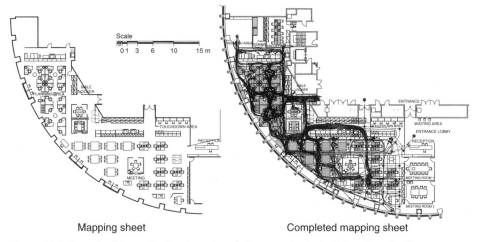

Mapping sheet   Completed mapping sheet

Figure 14.1 Example of data collection through workplace mapping in the Kokuyo branch office, Nagoya, Japan.

the work area and indicates all desks, filing cabinets, meeting spaces, and any other relevant support spaces (Figure 14.1). The various mapping procedures (activity, activity duration, movement, and communication) are carried out simultaneously at certain intervals. In the case study described below, 15-minute intervals with a 5-minute break between each interval were used. This break period is regarded as necessary to allow the surveyor adequate time to prepare for the next mapping period. During mapping periods, surveyors record the following information on the mapping sheet:

- time of the mapping interval;
- the specific department;
- the surveyor's name;
- all workplace activities – these are to be indicated according to the previously agreed set of activity abbreviations, together with starting and end time;
- the mode of communication (formal, informal, etc.), together with the location of each point of communication;
- all movement lines for each individual employee.

The complementary use of workplace mappings and employee surveys has various advantages over using questionnaires or interviews alone. Questionnaires and interviews provide valuable insight into employees' personal opinions of the workplace environment, but fail to provide objective information regarding the nature of employees' workplace activities or communication characteristics. Experience from data analysis of various BPE studies suggests that findings from self-reporting by employees regarding their personal work style, communication, and movement characteristics, are different from data collected through workplace mapping. This is particularly true with regards to questions about co-workers they have contact with, the frequency and duration of such contacts, and personal work style. The use of workplace mapping provides the researcher with more objective and comprehensive data on the nature of workplace activities and employee behaviour. Data obtained through workplace mapping also enables the researcher to relate the findings from a particular workplace environment to BPE performance levels more accurately.

However, the workplace mapping methodology has some inherent weaknesses. For example, individual interpretation of the mapping methodology by team members from different organizations was found to result in various problems. These problems included: differing mapping styles which caused confusion; members of the research team emphasizing different aspects of the mapping survey based on their respective research interests; and poor coordination between members of the research team surveying different parts of the workplace, since it is usually divided into smaller segments for ease of mapping. Another problem pertains to members of the research team memorizing and recognizing employees in their specific workplace mapping area. It is very important for the surveyor to identify correctly which employees are accommodated within his specific mapping area, since failure to do so can result in incorrect data being collected. Unless they are controlled, these problems can result in incorrect data that do not reflect the true nature of the various groups in the workplace environment. Effective control and administration of the research effort, including training all members of the team to have a correct understanding and application of the research methodology, is therefore essential to ensure successful performance evaluation.

14.4   Case study: workplace mapping in an innovative workplace

Based upon its location in a newly completed office tower in Nagoya, Japan, the Kokuyo branch office was selected to illustrate workplace mapping. This workspace is designed according to alternative workplace planning principles (for example, non-territorial workstations), and has achieved a high success rate in terms of the NOPA minimum standard. Prior to the research team's involvement, the company's management introduced a new and innovative workplace layout based on the spatial requirements of various workplace practices and processes, information and communication technology requirements, and the classification and formulation of staffing typologies according to work style diversity (see Figure 14.2). The study was designed and conducted with the specific purpose of identifying and analysing the effects of these new planning and staffing typologies on users' perceptions. The research methodology was designed to collect and benchmark data on behavioural aspects of Japanese work environments.

A main objective of the study was to evaluate staffing typologies in terms of their effectiveness by using workplace mapping in conjunction with data from a questionnaire distributed to all employees. During workplace mapping, observations were made on the basis of individual movement characteristics and interpersonal contact between employees. The workplace was divided into a series of nodes and paths, each of which was numbered and the frequency of employees passing through it registered. This enabled the research team to construct a model of communication behaviour in the workplace environment (see Figure 14.3).

Utilization studies of space and time resulted from observing and mapping employees' behaviour at work, occupancy of individual workspace, and movement through the space. Data from video footage were also analysed. This process of checking and comparing information from various sources increases reliability of results, and provides a better perspective on complex behaviour in the workplace environment.

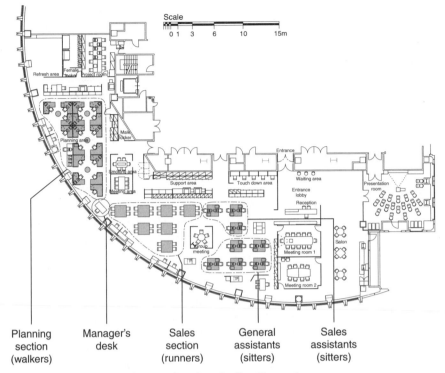

Figure 14.2 General workplace layout, Kokuyo branch office, Nagoya, Japan.

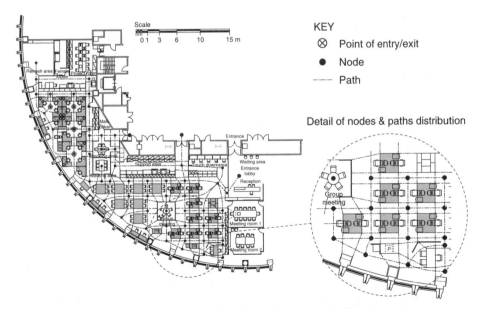

Figure 14.3 Path-and-node model of the workplace layout, Kokuyo branch office, Nagoya, Japan.

14.5   Staffing typologies

'Sitters', 'walkers', and 'runners' (see Figure 14.2) characterize employee work styles in the area observed. 'Sitters' are mostly administrative personnel who spend all their time in one workplace and are accommodated through a fixed-address system. Workspace for sitters comprises a 1600×1600 mm table accommodating two employees, each with an 800×1600 mm-sized desk space. In addition to this, sitters also have individual desktop telephones and personal computers, chairs with maximum support and arm rests, and a mobile filing cabinet. A partition between the two employees at each table ensures personal workspace, and provides some degree of visual screening and opportunity for concentrated work.

'Walkers' have a higher degree of mobility in the workplace; they include, for example, staff from the Planning and Design section. Their activities are focused on group work, and they are accommodated in groups of four, separated from adjacent groups by low partitions. These partitions provide group identity, but do not physically separate the enclosed group from the rest of the work environment.

Employees from the Sales section, who spend relatively little time in the office, are categorized as 'runners'. Due to their high level of mobility in the office, they are accommodated in a non-territorial or 'free-address' system; 1600×1600 mm worktables are used to accommodate a maximum of four employees. All connections for power and data are at table-top level to ensure ease of access, use, and functionality. Runners are also dependent on mobile (cellular and PHS) communication. Free-address workspaces have chairs without arm rests and mobile filing cabinets, which are 'parked' underneath individual lockers. Stand-up counter tops with filing and fax facilities are provided for use by runners, who spend a short time in the office each morning, and can meet informally while sending and/or receiving faxes.

14.6   Workplace mapping results

Data from workplace mapping were analysed to determine movement and communication patterns. By employing the node-and-path model, it was possible to determine movement and behaviour patterns for selected (key) employees. This process involves three types of analysis. The first, illustrated in Figure 14.4a, is a 'communication wheel': a circular diagram on which all employees are recorded according to their departments. Lines of varying thicknesses are used to connect employees. This method is useful in illustrating the characteristics of communication between employees, as the various line widths indicate frequency of communication between individuals, as well as communication between various departments.

The second type of analysis illustrates employees' patterns of circulation and route selection in performing their workplace activities (see Figure 14.4b). Again, lines of varying widths indicate the number of times the same route is used. Circulation patterns were constructed for employees from all three departments in order to illustrate differences and similarities in movement behaviour.

A third type of analysis focuses specifically on characteristics pertaining to employees' selection of movement routes in the performance of work-related tasks, as well as the various locations where they are performed (see Figure 14.4c). For example, Employee 1 (E1) conducts about 60 per cent of his tasks while moving between various points in the

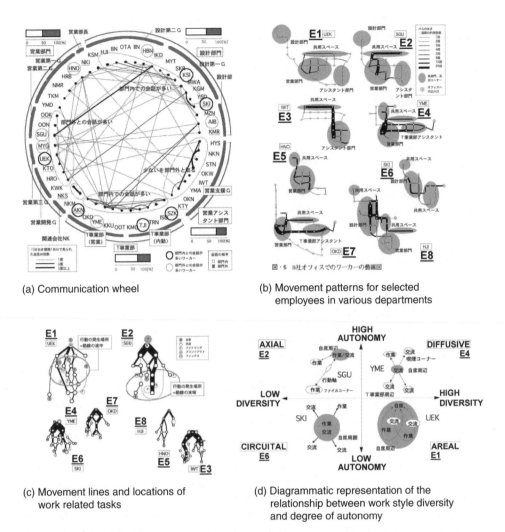

(a) Communication wheel

(b) Movement patterns for selected employees in various departments

(c) Movement lines and locations of work related tasks

(d) Diagrammatic representation of the relationship between work style diversity and degree of autonomy

Figure 14.4a–d Results of workplace mapping data analysis.

office. In contrast to this, Employee 2 (E2) carries out his workplace activities at the end of certain routes. Thus, E1 engages in workplace activities in a successive or work flow manner, while E2 conducts workplace tasks in a more purpose-oriented manner. Although Figure 14.4c shows the result of a 3-hour activity duration study, these two tendencies are also found in the 15-minute work sampling studies.

Based on degree of work style diversity and degree of autonomy, work styles are categorized as 'axial' (low work style diversity and high autonomy), 'circuital' (low level of work style diversity and low autonomy), 'areal' (high work style diversity and low autonomy), and 'diffusive' (high work style diversity and high autonomy) (see Figure 14.4d). These work style classifications can be related to other interpretations of office work patterns (Duffy, 1997). Accordingly, an axial work style is considered the intermittent occupation of individual or shared workspace for concentrated study, in which movement within the workplace is focused on the fulfilment of needs not provided for by workspace designed to accommodate a complex variety of workplace activities. A diffusive work style is characterized by the performance of work in various locations depending on the specific requirements thereof. Due to highly diverse work, employees with a diffusive work style utilize a greater variety of workspace settings, thereby displaying the highly autonomous nature of their specific work styles.

This categorization, together with circulation and communication analysis, enables the formulation of departmental movement and communication profiles, which improves management's understanding of employee work styles and behaviour. These profiles can be used when designing workplace layout changes to accommodate employee needs and diversity. This categorization of work style diversity (see Figure 14.4d) liberates design professionals from traditional space standards and suggests a more work-pattern-based allocation of workspace. Based on current data, the non-territorial office responds to the requirements of employees with high levels of both work style diversity and autonomy. Similarly, characteristics of work style and autonomy in the other three staffing typology categories can be used to address employees' workplace requirements more directly.

14.7 Conclusions

Building performance evaluation through workplace mapping adds to the already multi-faceted approach of BPE methodology. Experience from surveys where workplace mapping has been used, and especially in cases where follow-up surveys were conducted, has proved the validity of this approach. Since the Japanese context values more objective measurements, data derived from this type of workplace analysis are expected to enjoy broad applicability.

Analysis of data collected using workplace mapping is time-consuming due to the large volume of data collected. However, since all the various mapping procedures are done simultaneously, the overall effort is more effective than separate investigations to gather the same data. Moreover, workplace mapping provides the researcher with direct insight into the culture of an organization, since it reveals movement and communication behaviours, thereby facilitating the interpretation and diagnosis of problem areas. As such, it would seem that workplace mapping is a tool that could usefully be applied to Phase 2 of building performance evaluation, that is, programming, as well as to the feedback loop of Phase 5 (occupancy), that is, post-occupancy evaluation.

The focus of workplace mapping is specifically on what employees do at work, and on how long they engage in various workplace activities, such as computer work, paperwork, telephone conversations, formal meetings, informal communication, filing, printing, fax/copy, and general movement. Data analysis addresses functionality, efficiency, and work flow, while behavioural measures focus on social, psychological, and cultural dimensions. The results of these efforts allow concrete conclusions to be drawn regarding a given workplace environment's building performance and level of quality.

Acknowledgements

NOPA serves primarily as a core organization for the promotion of new office environments in Japan, while JFMA is the functional steering organization for the promotion of

FM-related issues in Japan. Educational matters related to FM are overseen jointly by JFMA, NOPA, and BELCA (Building and Equipment Life Cycle Association), while FM certificate testing is managed only by JFMA (NOPA, 2003).

References

Becker, F. (1990). *The Total Workplace: Facilities Management and the Elastic Organization.* Van Nostrand Reinhold (Japanese translation by A. Kato, 1992).

Becker, F. and Steele, F. (1995). *Workplace by Design: Mapping the High-Performance Workscape.* Jossey-Bass Publishers.

Duffy, F. (1997). *The New Office.* Conran-Octopus Ltd.

JFMA (2003). Available from: http://www.fis.jfma.or.jp/fis/front/index.cfm (Accessed 15 August 2003) (Japanese).

Le Roux, P.C., Kato, A., Taniguchi, G. and Tsunekawa, K. (2001). Staffing Typologies and Placemaking: Workplace planning methodologies in the new Kokuyo Nagoya Office. In *Architectural Institute of Japan Tokai Branch Technical Research Report*, **39**, 657–660.

Mori, A., Tsunekawa, K., Kato, A. and Le Roux, P.C. (2002). A Study on relationships among plan compositions, work styles, and communication behaviours in the office. In *Architectural Institute of Japan Journal of Architecture, Planning and Environmental Engineering*, **551**, 123–127 (Japanese).

NOPA (2003). Available from: http://www.nopa.or.jp (Accessed 15 August 2003) (Japanese).

Preiser, W. (2001). Towards Universal Design Evaluation. In *Universal Design Handbook* (W. Preiser and E. Ostroff, eds) pp. 9.1–9.18. McGraw-Hill.

Preiser, W. and Schramm, U. (1997). Building Performance Evaluation. In *Time-Saver Standards: Part 1, Architectural Fundamentals* (D. Watson, M. Crosbie and J. Hancock-Callender, eds) pp. 232–238. McGraw-Hill.

15

Evaluation of innovative workplace design in the Netherlands

Shauna Mallory-Hill, Theo J.M. van der Voordt, and Anne van Dortmont

Editorial comment

Over the last decade many businesses have been engaged in making organizational changes, and adopting new management styles and ways of working. Concurrently, there has been a rise in the number of non-territorial 'flexible' office designs, based on job functions and work processes rather than on individually assigned workstations. Office buildings are becoming more 'intelligent' through the use of advanced building management systems, automatic indoor climate controls, innovative (day)lighting systems, and the like. Such innovations in workplace design are intended to facilitate organizational change, improve user satisfaction, increase efficiency, and lower costs. To cope with the rapid innovation and changing nature of work environments, environment-behaviour researchers in the Netherlands used the building performance evaluation and POE (post-occupancy evaluation) approach to measure workplace performance.

15.1   Overview

This chapter describes how several researchers in the Netherlands are using Building performance evaluation (BPE) to test innovative designs for workplaces. After summarizing the general background, drivers and objectives of BPE in the Netherlands, two case studies are introduced. The first is an effectiveness review and post-occupancy evaluation (POE) of a new, non-territorial office design for ABN-AMRO bank. The second focuses on the pre- and post-occupancy evaluation of a new 'intelligent' (day)lighting system for the Rijnland Water Board building. A number of observations and recommendations about BPE based on the case studies are included at the end of the chapter.

15.2   Building performance evaluation in the Netherlands

In the Netherlands, it is unusual to undertake performance evaluations throughout the delivery and life cycle of a building. POE, and all other phases of the integrative framework for BPE, as described by Preiser and Schramm (1997) and in Chapter 2, are not explicitly included in standard agreements between clients and their designers, consultants and contractors. BPE does, however, have a role in the Netherlands.

Office workers in the Netherlands have come to expect a relatively high quality work environment. The Dutch follow the decision-making tradition of the 'polder model', where consensus is reached through consultation rather than through top-down authority (van Riet, 2001). As a result of this culture of consultation-based processes, the regulations concerning the health and welfare of office workers in the Netherlands – known as the ARBO-besluit – are quite strict. Though not always comprehensive in nature, most building design delivery processes involve an effectiveness review, a programme review and a design review. Once a building is occupied, large organizations commonly employ health and safety officers, and even physicians, to deal with worker concerns and complaints.

The current trend towards organizational change and new ways of working has fuelled the demand for new and innovative workplace solutions. Before investing in such solutions, stakeholders want assurances that proposed innovative designs and building systems will meet the needs of their users.

15.3   Evaluation methods and performance criteria

In most POEs in the Netherlands, the main objectives are to test whether clients' goals and objectives have been achieved, and to improve the understanding of the complex relationships between facilities management, employee satisfaction and organizational goals and needs (Table 15.1). As will be demonstrated in the ABN-AMRO case study in Section 15.4,

Table 15.1 Goals and objectives of POEs of innovative offices

- To support the choice of a future office concept with a feasibility study (pre-occupancy evaluation)
- To be able to write a sound strategic brief and project brief (pre-occupancy evaluation)
- To test if clients' goals and objectives have been reached
- To record unanticipated results, positive or negative
- To improve the understanding of complex relationships between facilities and ways of working, organizational needs and user preferences
- To legitimize a continuation or adaptation of accommodation policies
- To steer improvement and upgrading of buildings
- To monitor trends and developments within office organizations
- To develop theories and tools to support complex decision-making processes
- Input for a database of office buildings, including best practices and worst cases to build up a body of knowledge and data for theory development and benchmarking.

Source: Volker and van der Voordt, 2004

Table 15.2 Common criteria measured in POEs of innovative offices

Frequently measured	Less frequently measured
• Employees' characteristics (gender, age, education, occupation) • Characteristics of working processes (what are people actually doing, when, where) • Characteristics of old and new workplaces (location, layout, desk sharing or not) • User satisfaction on accessibility of colleagues (physically, by phone or email), communication, concentration, privacy, thermal comfort, use and experience of facilities • Most positive and most negative aspects • Overall satisfaction • Perceived productivity • Critical factors in successful implementation and management of building-in-use	• Occupancy level • Actual behaviour (e.g. frequency of desk rotating, claiming a favourite desk) • Psychological aspects such as status, territoriality, social contacts and personalization • Organization's characteristics such as strategy, corporate culture, vision of the future • Employees' health and safety • Image, i.e. effects on attracting and retaining employees and clients • Actual productivity • Economic value added • Facility costs • Adaptability and future value

Source: Volker and van der Voordt, 2004

the typical focus of the investigation is on user satisfaction and organizational performance, Technical aspects and facility costs are included less often (Table 15.2).

Detailed evaluation of environmental systems in the workplace, such as heating and lighting systems, usually takes place only when a particular problem has been identified by a previous study or worker complaints. Environmental system evaluations tend to be diagnostic in nature, focusing on specific performance aspects, such as glare, user control, maintenance, and so forth. In some situations, such as the Rijnland Water Board building discussed later in this chapter, where new and relatively unknown innovative systems are being considered for use in the workplace, full-scale mock-ups are built to evaluate them before they are installed.

15.4   Case one: office innovation at ABN-AMRO bank in Breda

Today, many business units of ABN-AMRO bank apply innovative strategies for flexible working. These include the short- and long-term time-sharing of spaces such as: open plan or group offices for communication and routine work; cockpits for concentration; coffee corners for breaks and informal meetings; formal meeting rooms; and touch-down places for short-time activities such as reading mail and checking email. With new workplace concepts such as these, ABN-AMRO wants to support organizational and cultural change that is open and dynamic, and to improve its overall performance without decreasing occupant satisfaction.

The first large-scale flexible office built for ABN-AMRO was their regional office in Breda (Figure 15.1). The original office building was functionally and technically out of date and had to be renovated. There were two options: keeping individually-assigned

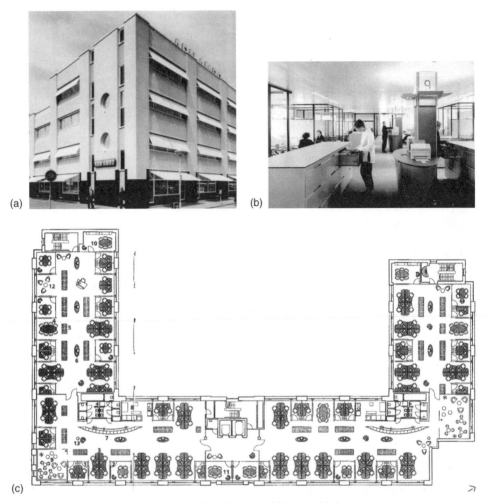

(a)   (b)   (c)

Figure 15.1 ABN-AMRO new innovative office, (a) exterior, (b) interior, (c) plan.

desks and extending the building by $2\,600\,\text{m}^2$; or keeping the same floor area and intro-ducing desk sharing. A cost comparison showed the desk sharing option increased invest-ment costs by 9 per cent, and reduced operation costs per employee by 17 per cent (Lohman and van der Voordt, 2000). Given this potential cost reduction, combined with a space-saving of nearly 30 per cent, the managers adopted the innovative option, referred to as the 'Flexido-concept'.

In the new configuration, 336 employees share 255 desks. This consists of 311 full-time employees using 194 desks in an open layout; 61 'cockpits'; 17 'touch down' workplaces; 18 team rooms; 15 meeting rooms; and 30 seats for informal collaboration, for a total of 400 places. A POE comparing user satisfaction pre- and post-move revealed the overall space reduction has not led to user dissatisfaction (van den Brink, 2000) (van der Voordt and Diemel, 2001). 69 per cent of occupants are positive about the layout, compared to 37 per cent in the previous open-plan office with individual desks. 51 per cent of users are positive about the effect of the physical environment on their productivity (formerly 14 per cent) and 83 per cent of people surveyed do not want to return to the former working environment.

As a consequence of flexible working, where employees may use different types of workplaces for long or short periods of time, people tended to plan more of their activities in advance, which has improved their effectiveness. By using team archives instead of personal archives, the space needed for filing was reduced by almost 50 per cent, with only a slight decrease in user satisfaction. Communication has improved slightly in the more open plan, and enclosed 'concentration cells' allow people to work in a concentrated way when necessary. The overall success of this project can also be attributed to a careful design delivery process that included an inspiring 'champion' from management to lead the process; sound pre-occupancy research into spatial needs; good communication with users; and post-occupancy care to solve minor problems.

According to pre- and post-occupancy evaluations in several other ABN-AMRO flexible workplace projects, most of the employees surveyed tend to be satisfied with the spatial transparency, aesthetic interior design, ergonomic furniture and the improved freedom of choice of when and where to work (van der Voordt and Beunder, 2001). Compared to cellular enclosed offices, improved transparency aids communication and, unlike in the previous open-plan offices, concentration is improved by adding 'cockpits' or small, enclosed individual workspaces. In time, people become accustomed to desk sharing and using different task-specific areas (desk rotating). However, users complained about distractions caused by a lack of visual and auditory privacy, time loss from repeated logging in, and the clean desk policy. Overall, the users' evaluation of the new workplace is slightly more positive than negative. In all the ABN-AMRO flexible workplace projects evaluated, the majority of employees surveyed say they do not want to go back to their former workspace.

15.5   Case two: building system innovation in Rijnland Water Board building

In 1998, the provincial government of Rijnland decided it needed a new building to house the Water Board. The new building needed to accommodate 350 employees. The two key performance requirements of the new facility were that it be highly energy-efficient (sustainable) and that it provide a supportive and comfortable working environment for its occupants.

To address the requirements for energy efficiency and indoor comfort, several innovative environmental-building systems were incorporated. These include hot and cold underground storage tanks, heat pumps, under-floor heating, radiant cooled ceilings, and a ventilation system that includes heat recovery. The most innovative measure was to install a new type of daylighting system.

As part of the overall effectiveness review, a three-part performance evaluation of the innovative lighting system was undertaken:

1. pre-testing of innovative daylighting designs in test settings;
2. evaluation of the selected system on site; and
3. pre-move and post-occupancy user surveys.

15.5.1   Pre-testing

The goal of the first phase of evaluation was to select one out of four innovative daylighting system solutions. Mock-ups of each daylighting design were created in existing offices at the Eindhoven University of Technology. Each room was set up to simulate the size, layout, and materials of the offices in the proposed new design for Rijnland.

Over a period of four months, data were collected and analysed for each system, according to the measured-observed-perceived-simulated (or MOPS) model of building environment evaluation (Mallory-Hill, 2004):

- *Measured* – monitoring physical performance through data loggers connected to lighting sensors, and spot measurements to determine light levels and energy use.
- *Observed* – walk-throughs and time utilization studies to track occupant activities and responses.
- *Perceived* – user surveys or interviews to capture occupants' satisfaction with the environment.
- *Simulated* – computer visualizations to examine the performance of the system year-round, and in different exterior weather conditions.

A photograph and plan of one of the experimental settings is shown in Figure 15.2. For a more detailed explanation of the evaluation see Zonneveldt and Mallory-Hill (1998) and Mallory-Hill (2004).

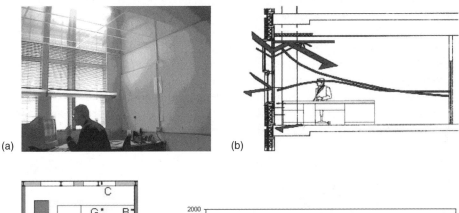

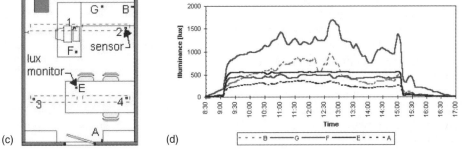

Figure 15.2 Evaluation of an innovative daylighting system: (a) design mock-up, (b) section of conceptual design (Bartenbach Lighting Laboratory), (c) and (d) physical measurements of lighting levels (Zonneveldt and Mallory-Hill, 1998).

The most successful system in terms of energy efficiency was selected by the client. The selected system is designed by Bartenbach 'Lighting Laboratory' in Aldrans, Austria (see Figure 15.2b). It combines indirect high efficiency electric lighting with reflective ceiling finishes and computer-operated reflective blinds, to optimize daylight penetration into the room. The concave profile of the venetian blinds further helps to 'scoop' light into the room. Luminaires near the back of the room are controlled by light sensors directed at the work-surface. When more daylight enters the back of the room, the electric lighting is automatically dimmed. The system was so effective in using natural light that no electrical lighting was required during 70 per cent of daytime working hours.

15.5.2   On-site testing

After the new building was constructed, further testing was undertaken to optimize the daylighting system design with employees from the Rijnland Water Board (van Wagenberg et al., 1998). A new test setting was set up on the ground floor of the new building. In total, 28 employees evaluated the system over two days during which they did their normal work in the test environment. Each participant completed a questionnaire about their general opinion of the daylight system, their preferred type of blinds (perforated or non-perforated), view to the outside, illumination level, and satisfaction with working in the new office.

The results showed that 64 per cent of the employees were satisfied with the light quality, compared to their normal worksetting where only 39 per cent were satisfied. When asked about the outside view (blinds down, half closed), 50 per cent of the subjects were dissatisfied. 40 per cent of participants preferred perforated over non-perforated blinds because of the improved view. This finding was lower than in their normal workstations, where 57 per cent of the subjects were satisfied with their view outside.

While working in the test setting, nearly half (47 per cent) of all of the subjects pulled up the blinds, both perforated and non-perforated, and a majority (67 per cent) switched on the light at the window, or the ceiling light, or both. These occupant modifications effectively compromised the majority of the energy-saving features of the design.

15.5.3   User surveys

Before and after the move, users were surveyed on how satisfied they were with the building regarding its location, building characteristics, and amenities such as catering, cleaning, layout, and maintenance. Users were also queried about workplace characteristics such as the indoor climate, effects on health and productivity, and the innovative daylighting system (van Wagenberg, 2001).

Most findings showed a significant increase in employee satisfaction in the new building (Figure 15.3), although moving from the historic centre of the city to the new location in a business park increased employees' dissatisfaction with location from 1.5 to 27.1 per cent.

Increases in satisfaction relate to indoor climate factors and perceived productivity. Formerly, 34.9 per cent of occupants complained about the overall indoor climate, compared to 19.9 per cent in the new facilities. Figure 15.4 shows the results for each indoor

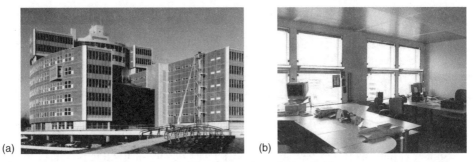

(a)   (b)

Figure 15.3 New Rijnland Water Board building: (a) exterior and (b) typical four-person workstation with innovative daylighting system.

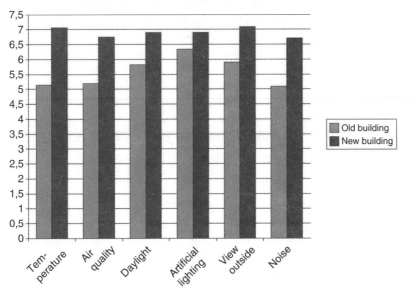

Figure 15.4 Pre- and post-occupancy evaluation of the Rijnland Water Board building (van Wagenberg, 2001).

climate factor. In the old facilities, all scores were below six on a scale from 1 to 10. In the new building, every indoor climate factor received an average score of seven.

The employees' self-rated productivity improved significantly. Previously, 42 per cent of the employees believed their building had a negative influence on their productivity, as compared to 9 per cent in the new building. Now, 33 per cent of the employees say the building environment has a positive influence on their productivity, compared to 3 per cent in the previous building.

15.5.4   Discussion of Rijnland Water Board results

In this study, performance evaluation in experimental settings helped to select, and then to optimize, an innovative daylighting system. Satisfaction with the innovative daylight system in the new building has increased, but not as much as expected. This partly relates to occupant expectations regarding blinds and the distribution of light. In the daylighting system that was selected, the blinds must be closed to work effectively; to most occupants, blinds are associated with keeping daylight out and not with bringing in extra light. However, when these blinds are closed, the illumination level produced by the system is evenly distributed in the room. The level of lighting is adequate, but some users find the lighting quality to be too uniform, or dislike the obstructed view outside. Occupants may, therefore, continue to open the blinds. Though this increases access to daylighting and view, it reduces the energy-saving performance of the system.

Similar comments from users about the perceived quality of lighting produced by the system were recorded in the very first experimental setting, but the substantial energy savings offered by this system made it very attractive to the client. The multidimensional evaluation methodology used in this study allows for the congruence between qualitative and quantitative results to be compared. Potential problems can occur, however, when quantitative performance measures, such as superior energy efficiency, are given priority over qualitative measures of occupant perceptions or observed behaviours, in the selection of the design solutions. The experience of this study suggests the primacy of occupant opinion should always be considered, even in limited, experimental settings.

15.6 Lessons learned from the two examples

Based on a comparison of the BPE methods used in the POE studies described in this chapter, the following observations and recommendations are made:

- Experimental settings are a good way to test innovative building systems. The challenge is to accurately replicate the proposed future setting. It is much harder to evaluate the success of flexible office design in purely experimental settings, as this is dependent on issues that are hard to replicate, such as organizational culture, emerging information technology, and outside economic or competitive pressures.
- User satisfaction is often used to judge the quality of physical environments for work, but is satisfaction a good measure of success? Many businesses are more interested in outcomes, such as improved productivity or cost-effectiveness. Workplace performance evaluation needs to find ways to measure the contribution of the space to business goals.
- There is a shortage of cost data in workplace performance evaluation. Surprisingly, most organizations do not record their cost data carefully. Ways to collect reliable data on investment, maintenance and running costs would be extremely valuable in supporting design decision-making.
- Many BPE/POE studies rely on user questionnaires alone. Using a variety of qualitative and quantitative measures allows congruence to be checked on different levels. Taken together, a multidimensional evaluation provides an accurate picture of overall performance, but the primacy of occupant opinion should be considered.
- Currently, workplace POEs are undertaken in a variety of ways. Standardized collection tools, definitions and classifications of performance measures would allow for the comparison and sharing of data and knowledge beyond individual case studies.

This chapter has shown that BPE plays an important role in helping to reduce risk, and promote understanding of the benefits and use of innovative design solutions for which few or no precedents exist. Flexible work patterns have spawned many innovations and changes in workspace layouts, furnishings, systems and equipment. Most employees, however, view workspace change with concern and suspicion: 'Are we downsizing?' 'What if I don't "get" this new technology?' The introduction of innovation into office design requires careful change management; changes need to be introduced both top-down and bottom-up, in a balanced way. In the Netherlands, building performance evaluation, combined with stakeholder participation throughout the design process, has helped to acquire and disseminate the information needed to help create better, more effective places to work.

References

Brink, A. van den (2000). *Flexido: de effecten in kaart.* Internal report: ABN AMRO & TU Delft. (Effects of the Flexido-concept on use and experience.)

Lohman, R.J.B.G. and Voordt, D.J.M. van der (2000). *Flexido: de kosten in kaart.* Internal report: ABN AMRO & TU Delft. (Cost implications of the Flexido-concept.)

Mallory-Hill, S. (2004). *Supporting strategic design of workplace environments with case-based reasoning.* Dissertation, Eindhoven University of Technology, The Netherlands.

Preiser, W. and Schramm, U. (1997). Building Performance Evaluation. In *Time Saver Standards* (D. Watson, M.J. Crosbie and J.H. Callender, eds). New York: McGraw-Hill.

Riet, K. van (2001, November). Letter from Holland: Polder-model design policy. *Doors of Perception Magazine.* Retrieved on 15 September 2003 from: http://www.doorsofperception.com/Features/details/10/?page=2.

Volker, L. and Voordt, D.J.M. van der (2004). *Werkomgevingsdiagnose-instrument.* In press. Center for People and Buildings. (Tool for a working environment diagnosis.)

Voordt, D.J.M. van der and Beunder, M. (2001). *De rode draad. Lessen uit innovatieve kantoor-projecten bij ABN AMRO.* Working document: Faculty of Architecture, Delft University of Technology. (Lessons learned from new offices.)

Voordt, D.J.M. van der and Diemel, L.H.M. (2001). Innovatief kantoorconcept bij ABN AMRO blijkt succesvol. *Facility Management Magazine* (**14**) No. 96, 34–42. (Innovative ABN AMRO office concept shown to be successful.)

Wagenberg, A. van (2001). *Evaluatie onderzoek Hoogheemraadschap van Rijnland Leiden.* Internal report: van Wagenberg Associates, Eindhoven, The Netherlands. (POE of the Waterboard of Rijnland, Leiden.)

Wagenberg, A. van et al. (1998). *Proefkameronderzoek hoogheemraadschap van Rijnland, Leiden.* Internal report: van Wagenberg Associates, Eindhoven, The Netherlands. (Mock-up research for the Waterboard of Rijnland, Leiden.)

Zonneveldt, L. and Mallory-Hill, S. (1998). Evaluation of daylight responsive control systems. In *Proceedings of the International Daylighting Conference '98.* Ottawa, Canada. 10–13 May 1998, pp. 223–30.

16

Evaluating universal design performance

Wolfgang F.E. Preiser

Editorial comment

In this chapter, the conceptual framework for building performance evaluation presented in Chapter 2 is extended to address universal design, the new, inclusive paradigm for design in the twenty-first century. The evolution of universal design is reviewed as a movement which seeks to make categories of daily necessities accessible and usable by a majority of people. Such categories might include products, interior architecture, buildings, urban spaces, public transportation, public parks (both urban and at the state/national levels), and information technology. The criteria for universal design performance were formulated by the Center for Universal Design at North Carolina State University in 1997, and are called the seven principles of universal design (Story, 2001). Each of the principles is an abstract ideal, for example, 'Equitable Use', which is then translated into subsets of practical design guidelines, creating significant overlap with traditional building performance criteria presented throughout this book. While there is experience with, and a growing body of knowledge on home design from a universal perspective, evaluation-based universal performance criteria for other building types are still rare, and thus, the accumulated knowledge is quite limited.

The argument is made that more case study evaluations on universal design need to be carried out for the purpose of knowledge building. Such evaluations are based on a methodology developed at the University of Cincinnati.

16.1 Introduction

With the publication of the *Universal design handbook* (Preiser and Ostroff, 2001), a new design paradigm for the twenty-first century was revealed in its full depth and breadth, and on a global basis. From design guidelines and standards to case study examples and information technology, this work evidences increasing acceptance of, and activity in, universal design. This chapter seeks to shed light on three major issues: first, the degree to which the integrative framework for building performance evaluation (BPE) presented in Chapter 2 can be transformed into a universal design evaluation (UDE) process model, in order to help implement universal design to its full intent and potential. Second, whether the concept of universal design can be applied to different cultures without losing its validity in attempting to accommodate most of the people most of the time. And third, developing universal design criteria for all common building types, some of which may already exist at the level of product and home design. However, for most building types, operationalizing the lofty ideals implied in the principles of universal design remains a daunting task and research agenda for years to come.

Various definitions for universal design have been offered, including that by the Center for Universal Design (Story, 2001). The meaning of 'universal' in this context is to make products and environments usable by a majority of people, regardless of gender, disability and health, ethnicity and race, size, or other characteristics (Mace, Hardie and Place, 1991). However, some argue that 'universal design' is an oxymoron. As Stephen Kurtz has observed: '... the designer is faced with a multitude of groups, often conflicting, who do not share common educational or class values, and who have little experience in major decision-making.' (Kurtz, 1976)

Previous attempts at designing products and environments for use by all did not necessarily meet with success. The Usonian house, designed by Frank Lloyd Wright in the 1950s, was to make affordable housing accessible to everybody. In reality, relatively few of these houses were built and they were somewhat difficult to live in, due to the very small sizes of bedrooms and kitchens. What features should a universally designed house have in an age where mass-produced goods can be individualized by seemingly endless choices? On one hand, current car production techniques demonstrate that the consumer is king, and the same production line can assemble cars with seemingly limitless variations. Feedback, feed-forward and control are the watchwords (Preiser, 2001) in a world of changing paradigms from mechanical to living systems, in which creativity, information and knowledge are the new currency (Petzinger, 1999). On the other hand, in the US housing market, banks, through their financing programmes, dictate the features and size a house must have in order to be saleable; and the 'cookie cutter' approach to housing design is pervasive. One could argue that this means it is universal in an odd, self-limiting way. The result is that only a tiny minority of houses are designed by architects, and only very few meet universal design criteria. Contemporary standards of home design may lead to controversy in the same sense as airplane seat design. As the present debate in the US airline industry illustrates, the 'one size fits all' approach to seating passengers discriminates against larger persons who, according to existing policies, are asked to pay for a second seat.

Similarly, the so-called 'Universal Plan', or footprint, for offices and workstation layouts is considered cost-effective by reducing the need to reconfigure when organizational departments move or change (Vischer et al., 2003). The question remains whether this approach makes office workers more productive, especially in light of the fact that it is diametrically opposed to the inclusive, non-discriminatory universal design paradigm advocated in this chapter.

16.2   Universal design

How universal is universal? In a homogeneous community and culture, it is possible to define and describe cultural norms and expectations for products, spaces and buildings.

However, in a world that is getting ever more diverse and globalized, the question has to be asked whether, and if so, how any one standard or set of criteria can universally meet everybody's expectations and needs.

Multinational corporations, for example, may develop intra-company standards for space allocation, work flow and processes, as well as for amenities for offices and production facilities on a worldwide basis (Brown, 1993). The same can be observed with global hotel chains, which provide specific levels of quality and amenities for the traveller. Often, such quality standards are expressed in rating systems, such as those found in Michelin and Mobil guides, which attempt to use objective criteria for the rating of facilities, services, amenities, and other features, such as beautiful scenery. Whether this statement is true and verifiable has recently been called into question, given the report that some evaluations are falsified and the testers paid off by big chefs.

Maruyama (1977) may have envisioned universal design when discussing design criteria for future, permanent settlements in outer space, with continuous, long-term occupancy by multicultural and international crews. The stress of living and working together in limited amounts of space over extended periods of time demonstrates the difficulty of 'design for all' when attempting to meet everybody's physical, social, psychological and cultural needs.

Serious issues of relativity and establishing priorities in universal design arise when dealing with different cultural contexts. Not only do space, lighting and other standards vary considerably across cultures in identical types of environments, such as office buildings, but also economic conditions, technological developments, and culture-specific customs and patterns of space utilization add to the complexity of this question. For example, in Japan, only about 65 square feet are allocated to each office worker, as opposed to about double that in the USA. Lighting standards in UK offices are lower than in the USA for identical tasks and functions. In Scandinavian countries, people dress more according to the season, thereby reducing heating and cooling demand in buildings.

16.2.1 Universal design criteria

About 12 years ago, the main terminal at JFK Airport serving the now defunct Trans World Airlines was adapted to provide access to persons with disabilities. Ramps covered the once majestic stairs, designed by architect Eero Saarinen (see Figure 16.1).

Who ended up using the ramps? Everybody with wheeled conveyances, i.e. mothers with strollers, airline personnel and passengers using luggage with wheels, people servicing vending machines, and others. This was an example of universal design retrofit.

The principles of universal design developed by the Center for Universal Design constitute ideals that need to be operationalized for use in the real world and in everyday design situations. Some products, such as 'Mr Good Grips' kitchen utensils and Fiskars scissors, have been developed to meet universal design needs. Much less has been done to accomplish this goal in building environments, such as homes, offices, schools, and transportation facilities. Despite the fact that universally designed homes are available (Young and Pace, 2001), there is continuing resistance in the design professions and the building industry to adopting the new paradigm, and to incorporating universal design criteria into home design. Universal design solutions are perceived to cost more, which is not the case. Designers and builders seem reluctant to add yet another layer of 'constraints' to their projects, in this case, universal design.

Figure 16.1 Ramps at JFK TWA Terminal.
Source: Wolfgang F.E. Preiser.

A number of building types and case studies have been published and serve as examples to be emulated (Preiser and Ostroff, 2001a). Strategies such as the three described below are needed to ensure that universal design criteria are operationalized and implemented on a broad scale.

1. short term: carry out (post-design) evaluations of a cross-section of existing building types, using the universal design evaluation (UDE) process model outlined below;
2. medium term: carry out (pre-design) programming projects on a range of future buildings by incorporating universal design criteria from the start, and by integrating them with existing standard building performance criteria;
3. long term: universal design education – infuse universal design into curricula of design schools as a required subject matter, in hopes that, ultimately, professionals will practice what they have been taught.

16.3   Universal design performance and building performance evaluation (BPE)

The goal of universal design is to achieve universal design performance of designs ranging from products and occupied buildings to transportation infrastructure and information technology. The principles of universal design referred to above constitute performance criteria, and they are the means whereby universal design aims to achieve four major objectives:

● it defines the degree of fit between individuals or groups and their environments, both natural and built;

- it refers to the attributes of products or environments that are perceived to support human activity in a way that is contextually appropriate;
- it aims to minimize adverse effects of products and environments on their users, such as: discomfort, stress, distraction, inefficiency, and sickness, as well as injury and death through accidents;
- it comprises not absolute, but relative, concepts, subject to different interpretations in different cultures and economies, as well as temporal and social contexts; thus, it may be perceived differently over time by those who interact with it, such as occupants, management, maintenance personnel, and visitors.

Conceptually, the universal design evaluation (UDE) process model has evolved from the consumer feedback-driven, evolutionary evaluation process models described in Chapters 1 and 2, i.e. post-occupancy evaluation (POE) and building performance evaluation (BPE). The major difference between them is the universal design performance criteria, which are derived from the principles of universal design (see Appendix A). However, in interpreting the principles and applying them to concrete building types and situations, it is clear that there is significant overlap with the traditional, evolving performance criteria and levels of BPE.

16.3.1   Performance levels

Building performance evaluation can be structured according to three levels of performance criteria pertaining to user needs (see Chapter 1). At each level, goals might include safety, adequate space and spatial relationships of functionally-related areas, privacy, sensory stimulation, and aesthetic appeal. For a number of goals, performance levels may interact and also conflict with each other, requiring resolution.

For each setting and occupant group, respective performance levels for pertinent sensory environments and quality performance criteria need to be specified. The three performance levels are:

Level 1   Health, safety, and security performance;
Level 2   Functional, efficiency, and work flow performance; and,
Level 3   Psychological/social, and cultural performance.

These three performance levels correlate with codes, standards and guidelines that designers can use. Level 1 pertains to building codes, and life safety standards projects must comply with. Level 2 refers to the state-of-the-art knowledge about products, building types, and so forth, exemplified by agency-specific design guides or reference works, such as *Time-Saver Standards: Architectural Design Data* (Watson, D., Crosbie, M.J. and Callender, J.H. (eds) 1997). Level 3 pertains to research-based design guidelines, which are less codified but nevertheless of importance for building designers and occupants alike. For example, research-based performance criteria have been developed at the University of Wisconsin-Milwaukee for such building types as Alzheimer's centres and pre-school environments (Cohen and Weisman, 1991).

The relationships and correspondence between evolving performance criteria (see Figure 1.2) and the principles of universal design are shown in Figure 16.2.

Investigating the overlaps and discrepancies between traditional and universal performance criteria presents opportunities for funded research in the future. As with any type of applied, cumulative knowledge, it will take a great number of universal design evaluation

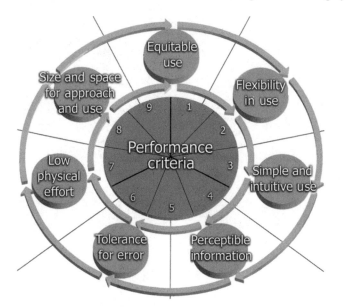

Figure 16.2 Universal design principles versus performance criteria (Figure 1.2).
Source: Jay Yocis, University of Cincinnati.

case studies to build up databases and clearinghouses that architects and designers can use. To start with, the most likely places where this research will take place are academic institutions.

16.4   Universal design evaluation (UDE)

Baird et al. have showcased a variety of building evaluation techniques, many of which lend themselves to adaptation for purposes of universal design evaluation (Baird et al., 1996). In that same volume, this author presented a methodology for three-day training workshops and prototype testing of medical office buildings, which involved both the facility planners/designers and building occupants (Preiser, 1996). This format, when adapted to universal design evaluation, could be very effective in generating useful performance feedback data for future building design. Figure 16.3 shows the universal design evaluation (UDE) process model, presented in terms of the conceptual frameworks outlined in Chapters 1 and 2.

The major benefits and uses of the universal design evaluation process, when applied to UDE, include:

- identifying problems in building performance and developing design solutions from an inclusive, universal design perspective;
- learning about the impact of practice on universal design and its impact on building occupants in general;
- developing guidelines and performance criteria for enhanced universal design concepts and features in buildings, and in the built environment in general;
- creating greater public awareness of the successes and failures of universal design.

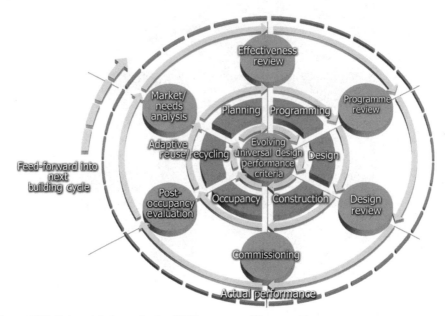

Figure 16.3 Universal design evaluation (UDE) process model.
Source: Jay Yocis, University of Cincinnati.

It is critical to formalize and document the expected universal design performance of new or retrofitted facilities in the form of qualitative criteria and quantitative guidelines and standards.

16.5   Strategies for universal design evaluation

In post-occupancy and building performance evaluation, it is customary to include Americans with Disabilities Act (ADA) standards for accessible design. The ADA standards provide information on compliance with prescriptive technical standards, but say nothing about performance, i.e. how the building or setting actually works for a range of users. Moreover, the principles of universal design also constitute a set of performance criteria and guidelines oriented to occupants' needs.

Case study examples of different building types, which incorporate universal design, are needed to advance the case for universal design in industry. These include living and working environments, public places, transportation facilities and recreational and tourist sites, among others. The case studies would use standardized data-gathering methods such as interviews, surveys, direct observation, photography, and videotaped walk-throughs of different facility types, for various groups of users. The resulting universal design critiques would focus on the seven principles of universal design and their corresponding universal design performance criteria. Other UDE examples pertaining to the design of bathrooms and kitchens are currently in development through the Rehabilitation Engineering and Research Center at the State University of New York at Buffalo.

At the University of Cincinnati, the author has guided a number of UDE case studies based on the principles of universal design and their common sense interpretations.

Figure 16.4 Hockey arena designed in the spirit of universal design.
Source: Wolfgang F.E. Preiser.

Systematic walk-throughs are conducted by students, covering all spaces of a building, including outdoor areas. A tabular, checklist format is used to record both positive and negative performance aspects, resulting in an executive summary with recommendations for problem resolution. To date, projects of this type have dealt with a hockey/sports arena (see Figure 16.4), a major airport, Alzheimer's centres, an (historic) adaptive reuse assisted living facility, state park service facilities, campus greens at a university, and a temporary dining hall structure, as well as other building types.

Additional case studies will be carried out and documented electronically while students are working for architectural and interior design firms. These will be deposited in an archive of the Center for the Study of Practice and will serve as precedents for future, similar research.

Methodologically appropriate ways of gathering data from populations with different levels of literacy, education and culture (Preiser and Schramm, 2002) are needed. It is hypothesized that through these methodologies, culturally and contextually relevant universal design criteria will be developed over time. This argument was eloquently presented by Balaram (2001) when discussing universal design in the context of an industrializing nation, such as India. The occupant as 'user/expert' (Ostroff, 2001) also has a role to play. The process of user involvement is often cited as central to successful universal design, but has not been systematically evaluated in regards to its effectiveness.

16.6   Conclusions

For universal design to become viable and truly integrated into the building delivery cycle of mainstream architecture and the construction industry, it will be critical to have all future students in building-related fields familiarized with universal design. It is also necessary to demonstrate to practicing professionals, the viability of the concept through a range of UDEs, including exemplary case study examples.

This chapter has attempted to demonstrate the logical evolution from BPE to UDE, by linking universal design performance to feedback, feed-forward and control of the building delivery and life cycles. Now that the 'performance concept' and universal design criteria (Story, 2003) can be made explicit and scrutinized through UDEs, they are gradually becoming an accepted part of a good design by moving from primarily subjective, experience-based evaluations to more objective evaluations based on explicitly stated universal design performance criteria in buildings.

Critical to the notion of universal design evaluation is the focus on the quality of the built environment as perceived by its occupants. In other words, building performance is not limited to energy conservation, life cycle costing, and the functionality of buildings; it also needs to focus on users' perceptions of buildings.

The cost of implementing universal design has frequently been cited as unjustifiable. However, it has been demonstrated that when integrated into building design from the outset, universal design features need not cost more than standard construction. Moreover, UDEs have become more cost-effective due to the fact that shortcut methods have been devised that allow the researcher, or evaluator, to obtain valid and useful information in a much shorter time-frame than was previously possible. Thus, the costs of staffing evaluation efforts and other expenses have been considerably reduced, making UDEs widely affordable. Universal design is a logical next step from the building performance evaluation process, and the implementation of a rational and research-based design process promises to create buildings that are, above all, good for the people that occupy them.

Acknowledgements

Figures 16.1 and 16.4 were prepared by the author, and Figures 16.2 and 16.3 by Jay Yocis at the University of Cincinnati.

References

Baird, G. et al. (eds) (1996). *Building Evaluation Techniques*. McGraw-Hill.

Balaram, S. (2001). Universal Design and the Majority World. In *Universal Design Handbook* (W.F.E. Preiser and E. Ostroff, eds). McGraw-Hill.

Brown, J.M. (1993). NCR Corporation. Worldwide Facilities Design Guidelines. In *Professional Practice in Facility Programming* (W.F.E. Preiser, ed.). Van Nostrand Reinhold.

Center for Universal Design (1997). *The Principles of Universal Design* (Version 2.0). North Carolina State University.

Cohen, U. and Weisman, G.D. (1991). *Holding on to Home: Designing Environments for People with Dementia*. The Johns Hopkins University Press.

Kurtz, S. (1976). Nothing Works Best. *Village Voice*, August 2.

Mace, R., Hardie, G. and Place, J. (1991). Accessible Environments: Toward Universal Design. In *Design Intervention: Toward a More Humane Architecture* (W.F.E. Preiser, J.C. Vischer and E.T. White, eds). Van Nostrand Reinhold.

Maruyama, M. (1977). Human Needs in Space – Psychological and Cultural Considerations. In *Space Settlements – A Design Study* (R.D. Johnson and C. Holbrow, eds), National Aeronautics and Space Administration, pp. 29–32.

Ostroff, E. (2001). Universal Design Practice in the United States. In *Universal Design Handbook* (W.F.E. Preiser and E. Ostroff, eds). McGraw-Hill.

Petzinger, T. (1999). A New Model for the Nature of Business: It's Alive. *The Wall Street Journal* (February 26).

Preiser, W.F.E. (1996). POE Training Workshop and Prototype Testing at the Kaiser-Permanente Medical Office Building in Mission Viejo, California, USA. In *Building Evaluation Techniques* (G. Baird, et al., eds). McGraw-Hill.

Preiser, W.F.E. (2001). Feedback, feed-forward and control: POE to the rescue. *Building Research and Information,* Vol. 29 (**6**), pp. 456–459.

Preiser, W.F.E. and Ostroff, E. (eds) (2001). *Universal Design Handbook*. McGraw-Hill.

Preiser, W.F.E. and Schramm, U. (1997). Building Performance Evaluation. In *Time-Saver Standards: Architectural Design Data* (D. Watson, M.J. Crosbie and J.H. Callender, eds). McGraw-Hill.

Preiser, W.F.E. and Schramm, U. (2002). Intelligent Office Building Performance Evaluation. *Facilities,* Vol. 20, No. 7/8, pp. 279–287.

Story, M.F. (2001). The Principles of Universal Design. In *Universal Design Handbook* (W.F.E. Preiser and E. Ostroff, eds). McGraw-Hill.

Story, M.F. (2003). Personal communication.

Vischer, J.C., McCuaig, A., Melillo, M. and Nadeau, N. (2003). *Mission impossible ou mission accomplie?: Évaluation du mobilier universel dans les édifices de Desjardins Sécurité Financière, 2 vols.* Montréal: Groupe de recherche sur les environnements de travail. Université de Montréal.

Watson, D., Crosbie, M.J. and Callendar, J.H. (eds) (1997). *Time-Saver Standards: Architectural Design Data.* McGraw-Hill (7th Edition).

Young, L. and Pace, R. (2001). The Next Generation Universal Home. In *Universal Design Handbook* (W.F.E. Preiser and E. Ostroff, eds). McGraw-Hill.

17

The facility performance evaluation working group

Craig Zimring, Fehmi Dogan, Dennis Dunne,
Cheryl Fuller, and Kevin Kampschroer

Editorial comment

Public agencies are under intense pressure to become more customer-driven and efficient, and to be clearer about performance goals and measurement of building performance. As a result, many public agencies have recently renewed their commitment to building performance evaluation (BPE) through a new collaborative effort.

The facilities performance evaluation (FPE) working group, a coalition of six large state and federal agencies, including California Department of General Services (DGS), US General Services Administration (GSA), Division of Capital Asset Management of Commonwealth of Massachusetts, Minnesota Office of State Architect, Naval Facilities Engineering Command (NAVFAC), and the Office of Energy Efficiency and Renewable Energy of Department of Energy (DOE), first met in 2002. The working group is sharing building evaluation experience, tools and results in an effort to reduce the cost and uncertainties of developing BPE programmes, and to be able to potentially benchmark the performance of their facilities against a much larger group of facilities. The agencies that constitute the working group produce over $10 billion in facilities each year.

17.1 Introduction

From the perspective of the BPE framework that guides this volume (see Chapter 2), the working group partners are aiming to increase the range of decisions that are informed by evaluation. The goal is to make evaluation a continuous process, starting immediately after occupancy, and impacting the full range of decisions from initial analysis of demand to final reuse or demolition. Unlike many private developers and clients, public agencies are usually responsible for the full cycle of building delivery, including demand-analysis, development, specification writing, master planning, schematic design, design, construction, operations, repair, renovation, and demolition or reuse. Public agencies often occupy the buildings they produce, or have long-term commitments to clients who do. As a result, these agencies have a demonstrable interest in how decisions about facility planning, design and operations affect long-term organizational effectiveness. Also, public agencies often have large inventories of buildings of similar functional types, such as offices, court-houses, hospitals or prisons, and can build up knowledge about them over time.

Indeed, public agencies have long been innovators in knowledge-based approaches to building delivery and management, and have been early adopters of the full cycle of the knowledge-based practices described in Chapters 1 and 2 of this book, such as performance-based specification writing, design guides, and design review. What is new in this effort is the working group's attempt to incorporate evaluation at different steps of facility delivery, and particularly to address those cultural, job, and process design and technical issues that have limited the effectiveness of previous evaluation programmes (Zimring and Rosenheck, 2001). The working group is seeking to create evidence-based and customer-based strategies for each BPE phase by:

● scanning internal, customer and industry trends and monitoring changes in work practices, living patterns, technology, social trends, and economics;
● setting performance goals;
● experimenting with innovative solutions;
● evaluating performance;
● interpreting findings and making them meaningful for action by different stakeholders;
● creating organizational memory;
● making results widely available, through databases, reports, meetings, internal experts, etc.;
● feeding-forward best practices into policy, project development and operations.

This chapter focuses on three current activities of the working group: building on lessons learned from previous evaluation programmes, especially those of the Disney Corporation; exploring development of comparable user survey tools; and attempting to create common energy metrics.

17.2   Lessons-learned from previous evaluation programmes

The initial efforts of the FPE working group were devoted to exploring past evaluation efforts and identifying strategies for success. Several of the senior executives have led sig-nificant previous evaluation efforts, and the consultants facilitating the group have a long history of conducting evaluations. Part of the strategy of the group is to identify 'best prac-tices'. The Disney Corporation is one such example; senior Disney executives cite their building evaluation programme as an important contributor to their overall success in developing facilities.

Disney projects are evaluated and developed by three groups working collaboratively: the imagineering, operations and construction management groups. The imagineering group is in charge of design and delivery for Disney projects and includes 3 000 'imagineers' in California, Florida, Tokyo and Hong Kong, coming from some 140 design, construction and management disciplines. The operations group, comprised mostly of industrial engineers, is responsible for programming the size and scope of new attractions. The construction managers oversee design and construction of projects.

At Disney, projects are guided by centralized corporate master architectural specifications that provide guidance on both design and operational aspects of all projects. These range from very detailed specifications for elements such as hotel reception counters and guest rooms, to broader, conceptual design guidelines. The master specifications include both design and programming elements, and detailed operational data, such as how long it takes to clean a guest room or paint a rail. The specifications are continuously updated by a wide variety of evaluation activities, such as observing customers, focus groups, surveys, and costs and performance monitoring.

Disney has maintained a commitment to evaluation since the 1970s. From the day they start work, all staff know that every new Disney project, ranging from a major theme park to a new show or meal, will be evaluated, and that the results will be openly discussed. Disney management teaches new personnel to overcome the potentially threatening nature of evaluation, and to learn to be objective in discussing the strengths and weaknesses of projects.

The evaluation practices that Disney has implemented cover a variety of issues related to the six phases of building delivery (Preiser and Schramm, 1997). They particularly emphasize the link between architectural programming, design, and operations, including routine maintenance. Disney provides an example for other agencies in establishing an integrated framework for performance evaluation. Perhaps most significantly, the company has created a data warehouse that includes over 30 years of evaluation data, which has become a valuable organizational resource.

Based on the experience of Disney and others, the working group has developed several principles and corollaries for creating sustainable evaluation programmes:

1. Create an organizational culture that supports evaluation and the use of evaluation results. Many organizations are what organizational theorists Argyris and Schon call 'self-sealing', where participants seal off performance results so that they are not held accountable (Argyris and Schon, 1996). Organizations with a culture that is able to embrace, or at least tolerate, open discussion of results without being threatened are best able to use evaluation. Therefore:
   1.1 include evaluation results and activities in a variety of venues, team meetings and discussions, internally and with clients;
   1.2 make evaluation part of a full range of formal and informal communications, websites and newsletters;
   1.3 infuse support for and participation in evaluation throughout the organization and at the senior executive, middle management and staff levels;
   1.4 base staff and project evaluations on empirically substantiated and broadly accepted criteria rather than subjective judgements.
2. Integrate ongoing evaluation into everyday work and processes. Successful organizations have been able to integrate evaluation into routine practices such as specification writing, guideline writing, project planning and facilities management. This integration was effective in:
   2.1 linking evaluation to the core mission of the organization. Whereas public sector organizations cannot use metrics such as profit margins, or even Disney's core question of the visitors' 'intention to return', public agencies have developed other measures of whether the facility is helping the organization achieve its core mission. GSA, PWGSC and others link facilities to their core mission using the balanced scorecard approach (Heerwagen, 2001). The balanced scorecard approach is a strategic planning methodology based on the idea that organizational

performance can only be adequately assessed by measuring outcomes on multiple categories and measures such as financial outcomes, internal business metrics, internal learning and growth and customers' satisfaction and behaviour (Kaplan and Norton, 1996);

2.2 ensuring that evaluation practices and results enhance other decisions and information. Few organizations have time or resources for large stand-alone evaluation programmes, but can justify evaluation if it demonstrably improves everyday operations;

2.3 making evaluation information available when desired and in a form that is useful. This requires that information be current and valid and fit into everyday decision-making. The recent development of information technology and Internet and intranet sites makes sharing information easier.

3. Develop a range of evaluation methods, including the ability to conduct rapid, inexpensive evaluations. Some previous evaluation efforts have depended on large-scale evaluations by consultants, but many agencies lack resources to do large evaluations on a routine basis. Agencies that have developed relatively low-cost and rapid methods, in addition to more targeted and larger-scale evaluations, are more likely to incorporate evaluation over the long term.

4. Develop baselines, benchmarks and comparisons. Public agencies have the potential to develop large data sets that allow them to compare the performance of single buildings within a particular portfolio of buildings, or of different portfolios of buildings against each other (Dillon and Vischer, 1988). But for such a comparison to occur, the data must be available and collected in parallel ways. There is, however, a potential tension between collecting common data that allows evaluation and comparison among sites, and evaluations that focus on the specific problems, needs and work processes of a specific site and organization. This is discussed in more depth in the conclusions section, below. Furthermore, public agencies often do not have the competitive concerns facing private sector firms, and are often able to make evaluation data available for further research and benchmarking by the private sector.

17.3   A common questionnaire

One of the goals of the working group is to develop more cost-effective and valid ways of collecting and sharing data. Building performance evaluations often include a user survey, and all the working group partners are committed to surveying the occupants of their facilities. Common or comparable items among the questionnaires used by the different agencies would allow the partners to build up larger sets of results and develop benchmarks more quickly than they are able to do individually. However, each agency has developed a unique survey based on its own history and goals, and pooling data is a complex problem. User responses can be strongly affected by minor changes in wording of questions (Dillman, 2000).

One of the authors analysed three user surveys used by the US General Services Administration (GSA), the Naval Facilities Engineering Command (NAVFAC) and Public Works and Government Services Canada (PWGSC), to understand possibilities for common development. The GSA client survey is an online survey prepared by the Center for the Built Environment at the University of California in Berkeley, as part of a set of three surveys administered to evaluate operational and managerial issues, design and construction issues, and occupant satisfaction levels. As part of their evaluation efforts, GSA has also developed an interface between their online surveys, an online database, and a reporting tool. The NAVFAC facility quality survey is prepared by the Engineering Innovation and Design Criteria office of NAVFAC. In its most current version, it is administered as an online survey, and is followed by focus group sessions with building occupants where survey results are discussed. NAVFAC has implemented an electronic database to store evaluation results, and an online reporting tool. Finally, the Public Works and Government Services Canada Employee Satisfaction survey is a brief, paper-based survey that is administered in early occupancy to identify any fine-tuning changes required. It is followed up by a similar survey after changes have been made, so that results can be compared.

The comparison of survey tools aimed at identifying how much overlap there is among them, and whether it is possible to create a larger database of evaluation results derived from current performance evaluation questionnaires. The working group hopes to:

1. identify existing common questions that could be compared, allowing pooling of data between agencies, or;
2. create new questions or modules that could become part of all surveys, or;
3. adapt one of the existing surveys for common use.

Table 17.1 compares the format, wording and content of questions in each of the three examples.

Table 17.1   The comparison categories used to analyse surveys

	Comparison categories	**Explanation**
Format	Language	Affirmative or negative Subjective personal pronoun
	Survey structure	Explicit thematic categorization in the form of differentiated survey sections
	Question styles	Drill-down Open-ended Yes/no questions Scale questions Multiple choice
	Scale	1 to 5, or 1 to 7
Content	Topics	User satisfaction related topics that are addressed by survey questions:

User satisfaction related topics that are addressed by survey questions:

1. Access	2. Orientation
3. Accessibility	4. Overall assessment
5. Acoustic	6. Parking
7. Air quality	8. Privacy (general)
9. Background Personal	10. Process
11. Building/grounds	12. Productivity
13. Design and colour	14. Recycling
15. Details	16. Safety/security
17. Ergonomics	18. Storage
19. Flexibility of space	20. Supporting spaces
21. Floor plan/layout/area	22. Survey comments
23. Furniture	24. Technology
25. Humidity	26. Temperature/thermal comfort
27. HVAC	28. Visual privacy
29. Light/natural and artificial	30. Wayfinding
31. Maintenance problems	

17.4   Results

The surveys vary in length and coverage. The PWGSC survey has 24 questions in total, with a single question for each of the 18 of 31 topics that it covers, except for several questions focusing on overall assessment and process. The NAVFAC survey has 94 questions covering 27 themes out of 31, with a strong emphasis on maintenance and on the delivery of new facilities. The GSA survey has 156 questions, or question items, with many 'drill-down' questions. (If a respondent marks that he or she is 'very dissatisfied', a follow-up question will pop up which asks for the rationale for the response.) However, because of the online administration of the survey, with automatic branching and drill-down questions, most users do not answer all questions and it appears much shorter to the respondent. The GSA online survey also allows development of modules for special users or building types, so that judges, for instance, do not receive questions aimed at security staff. The GSA survey includes questions related to 23 categories out of the 31, with a strong emphasis on accessibility, light, temperature, and wayfinding. Of the three surveys, the GSA survey has the potential to provide the most detailed information.

Only four topics are common to all three surveys: air quality; light; overall assessment; and office technology. In addition, the surveys differ in the phrasing of the questions and answers, use of scales and type of questions used, and the provision of an explicit thematic structure and survey instruction. These differences make results from the surveys difficult to compare.

This comparison offers several common directions for planning user surveys: the brief summary survey versus the longer in-depth survey; paper versus online surveys; surveys focusing on user experience with the operating building versus the process of project delivery. In Chapter 7, Bordass and Leaman describe a set of core survey questions that were tested in a number of building evaluations; and Chapter 1 also describes an example of a short set of survey questions aimed at evaluating building quality. Both tools are included in the Appendix at the end of this book. In addition, computer technology and modular survey design can potentially allow flexibility to adapt to different needs.

17.5   Developing methods and procedures for energy performance

The FPE working group is also establishing common metrics and procedures for energy performance. As with the user surveys, the large inventories of buildings managed by the partners provides a valuable opportunity to benchmark alternative design and management practices and to identify possibilities for improved energy performance. The working group has partnered with the National Renewable Energy Laboratory (NREL), who is conducting the Performance Metrics Project (PMP) for the Department of Energy. The PMP project seeks to define and standardize metrics and procedures to measure the energy performance of buildings, and to publicize these methods and procedures through partnerships with federal and state agencies. Some of the methods and techniques available are outlined in Chapter 9.

The complementary interests of the PMP and the FPE working group provide an opportunity to define, implement, and test procedures to ensure more consistent and reliable assessment, reporting, comparison, and benchmarking of energy efficiency. The collaboration between NREL and the FPE working group includes the following activities:

- Survey member agencies to gather information on current methods for measuring and tracking energy performance.
- Review methods and data for consistency, gaps, validity and reliability.
- Develop draft procedures for quantifying energy performance, including common definitions of terms, consistent methods for gathering and analysing data to support the metrics, and recommendations for reporting the results.
- Facilitate pilot tests of the procedures and metrics among member agencies, and make revisions on the basis of the results.

In addition, an 'energy tool kit' is being developed at The Georgia Institute of Technology. It measures the energy performance of a building, and tracks improvements or deterioration in efficiency over time. The tool kit uses the EPN (energy performance norm) calculation procedures, developed in the Netherlands (NEN 2916). The EPN was developed to estimate the building's energy use in a highly cost-effective, simplified procedure, with reasonable accuracy, without having to resort to computing-intensive dynamic simulation. The EPN calculates monthly and yearly energy consumption based on the breakdown of different energy consumers, such as for heating, cooling, lighting, humidification, fan power, and others.

Data regarding energy performance are often unique to a particular agency and unavailable to others. Even if data are made available, they often cannot be compared because they are collected, archived and analysed using different methods. When actual measurements are not available, energy performance assessments are often based on estimates that are not verifiable. In addition, uncertainties – for example, the percentage by which the calculated or measured value of energy may differ from the actual value – are seldom taken into consideration.

The logic behind any evaluation is to compare current conditions to other conditions. To be comparable the data need to be 'normalized', that is, mathematically adjusted to reflect systematic differences in building area, production unit, operational hours, weather, and operating conditions. But these adjustments are not always consistently applied. Other researchers have pointed out the necessity of using additional metrics in normalization and adjustment, such as the internal load density (lights, equipment, number of occupants) (Augenbroe and Park, 2002).

Comparisons and benchmarking in energy efficiency are difficult due to a lack of standard procedures and metrics, organizational issues, and differences in techniques for normalization. This makes it difficult to examine the performance of a building over its life cycle, to assess innovative solutions, compare different facilities or compare across agencies.

17.6   Discussion and conclusions

The members of the working group are attempting to address difficult organizational challenges by infusing better information and evaluation throughout their building delivery and management process. The organizations aim to reduce costs and organizational risk by sharing their experiences with building evaluation, and by sharing tools and results.

These activities show both the potential and challenge of collaborating on evaluation. The results have the potential to impact over $10 billion in annual construction, and the daily work life of some four million workers. However, each organization has its own culture, history and goals.

In addition, evaluation studies reveal a potential tension between standardization, and sensitivity to context. Whereas some evaluation studies are aimed at making comparisons among a range of buildings, others are more focused on understanding performance in a specific context that has a unique set of needs, goals and personnel. Zimring, Wineman and Carpman (1988) have argued that this is an inherent conflict in evaluation that is similar to conflicts in other kinds of social research: by recording the specifics necessary to understand the uniqueness of a case, the evaluator, by necessity, limits the generalization of the results.

The working group aims at overcoming this tension in several ways. It is attempting to create multiple models and forms of evaluation that include both standardized surveys and project-specific problem-solving. It hopes to pool large sets of evaluation results to allow analysis of trends across studies. As discussed above, a modular approach to building user evaluation surveys would allow comparability of data, while accommodating the needs of a specific setting.

Acknowledgements

This research has been funded by the US General Services Administration and the California Department of General Services. The energy section was based on a report prepared by Michael Deru and Joel Todd. Cheol-Soo Park contributed to parts of the energy section of this chapter.

References

Argyris, C. and Schon, D.A. (1996). *Organizational learning II: Theory, Method and Practice.* Addison-Wesley Publishing Co.

Augenbroe, G. and Park, C.S. (2002). *A Building performance tool kit for GSA.* Georgia Institute of Technology, Atlanta.

Dillman, D.A. (2000). *Mail and Internet surveys: the tailored design method* (2nd edn). New York: Wiley.

Dillon, R. and Vischer, J. (1988). *The Building-In-Use Assessment Methodology.* Ottawa, Canada: Public Works Canada (2 vols).

Heerwagen, J.H. (2001). *A Balanced Scorecard Approach to Post-Occupancy Evaluation: Using the Tools of Business to Evaluate Facilities.* Paper presented at the Federal Facilities Council Symposium on Building Performance Assessments: Current and Evolving Practices for Post-Occupancy Evaluation Programs, Washington, DC, March 13.

Kaplan, R.S. and Norton, D.P. (1996). *The balanced scorecard: translating strategy into action.* Harvard Business School Press.

Preiser, W.F.E. and Schramm, U. (1997). Building performance evaluation. In *Time-Saver Standards for Architectural Design Data* (7th edn) (D. Watson, M.J. Crosbie and J.H. Callender, eds). McGraw-Hill.

Zimring, C. and Rosenheck, T. (2001). *Getting it Right the Second or Third Time Rather than the Sixth or Seventh.* Paper presented at the Federal Facilities Council Symposium on Building Performance Assessments: Current and Evolving Practices for Post-Occupancy Evaluation Programs, Washington, DC, March, 13.

Zimring, C.M., Wineman, J.D. and Carpman, J.R. (1988). The new demand-driven post-occupancy evaluation. *Journal of Architectural and Planning Research*, December.

18

The human element in building performance evaluation

Alex K. Lam

Editorial comment

Building performance evaluation (BPE) consists of six phases – namely, planning, programming, design, construction, occupancy and adaptive reuse/recycling. The objective of BPE is to ensure both qualitative and quantitative criteria are met during design, construction and occupancy of a building.

A building is like any other object, such as a toaster, an automobile or a complex computer for the space shuttle. Based on a set of specifications, it can be constructed even to the most extreme tolerances. How well it will perform depends on how well the delivery process adheres to the specifications. Most evaluation systems are based on the technical performance in the delivery of the product. This process is crucial to ensure quality, adherence to the specifications and continuous improvement.

The building delivery system is a single integrative process, and it has a process leader; the six phases of BPE are sub-processes. Each of these sub-processes may also have its own process leader. In large and complex building projects, a single process leader is normally assigned to oversee the programme from start to finish.

This chapter deals with the pivotal role of the process leader in the building delivery process. We examine how the process leader must ensure accurate information is developed for the life of the project, and promote a participatory environment for collaborative work.

18.1   Introduction

The building delivery process begins with a problem, or need, and ends with a solution. How successful the solution is depends on how the need is communicated to the people who will devise the solution. This communication is in the form of instructions and specifications.

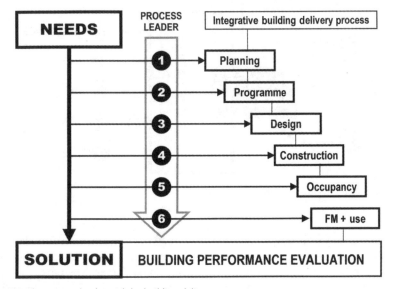

Figure 18.1 The process leader and the building delivery process.

The process generating the solution comprises the six phases of the building delivery system. Figure 18.1 shows the relationship among these phases. At each level of the process, the requirements of each of the six levels of sub-processes are communicated through the process leader. For each phase, the performance of each sub-process is evaluated. Since the six phases are interrelated, the entire evaluation system has to be integrative. The total system can then be evaluated as building performance evaluation.

Building performance is poor if the information provided to each sub-process is poor or inadequate. Each sub-process depends on the process leader to provide accurate and reliable information in order that a solution can be worked out. If, during the planning stage, the information regarding the purpose and use of the building is misleading, the planner will then produce a wrong set of site search requirements for the real estate professionals, incomplete design criteria for the architects and engineers, inadequate space accounting for the user community, and perhaps even an incorrect timetable to steer delivery of the project. Furthermore, this will affect the other five phases and the problem will be magnified as the project continues. The damages will be costly at the end.

Therefore, the role of the process leader, and of the communication of information, is of paramount importance to the success of the integrative building delivery process. Figure 18.1 shows that, at each phase, the process leader is involved and acts as the conductor of the flow of information.

18.2   The information generating process

In the integrative building assessment process, it is not difficult to realize that at each of the six phases, many professional parties are involved. At each phase, information and data must be generated in the form of specifications, needs analysis, and performance requirements (see Chapter 2). Communication comes in various forms – text, graphic or numeric. The

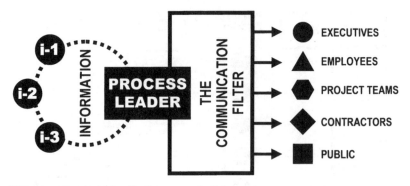

Figure 18.2 Integrative model for effective communication.

communication may be in a hard copy format or a soft copy or electronic format. However, this is only secondary; the primary source is the person generating the information. The persons who provide this information are the user communities. The person to organize and gather this information is the process leader. This person must work with the various parties to extract all the necessary information to make up the specifications for building design, construction and operations. As users do not always know exactly what they want, it is the job of the process leader to tease out the information and help them to identify and confirm their needs requirements (see Chapter 4).

Sydney J. Harris, the famous US journalist who wrote for *The Times*, once said, 'The two words "information" and "communication" are often used interchangeably, but they signify quite different things. Information is giving out; communication is getting through.' (Harris, 2003)

Figure 18.2 shows the integrative model for effective communication. The idea of information gathering is represented by the first stage of the diagram, with data coming in from various sources. The first task for a communicator is to ensure the accuracy and relevance of these data. The process leader must exercise precaution and discernment to sort through the information, such that only the supportive facts are used for communicative purposes. The contents of the communication must be appropriate for the intended audience.

18.3   The process leader

The cooperation between groups and information holders will depend on the ability and skills of the process leader. The success of the project depends a great deal on his or her personality and credibility. Special leadership and facilitating skills are required.

Warren Bennis suggests that the skill needed in leadership is the 'management of attention' (Bennis, 1999). He is referring to the idea of leadership quality in promoting contributions from the team. Along the same lines, Patrick J. McKenna emphasized that the 'first impression' is important in building teams and communicating to team players (McKenna and Maister, 2002). The leader/facilitator has to ritually welcome people into the team before the common goal is presented and shared. He suggested there is even a need to have pictures taken and post them on the wall of the team workroom.

In the corporate world, one always wants leaders to 'walk the talk', but this is not enough. Is the leader taking the lead to do what he or she advocates? 'Walk the talk' relates to the performance of the leader. What we need today is not only 'walk the talk' but also, more importantly, 'talk the walk'. To 'talk the walk' refers to credibility. It refers to the inner life – the ethics and morals of the leader's soul. How well communications are received depends on how trustworthy and credible the communicator is.

18.3.1   Qualities of a process leader

Australian author Mark Holden identifies that there are two elements in any manager's life, and they are performance and relationship. For performance, one needs to develop competencies and confidence. For relationship, one needs to develop and establish credibility and trust with the people. 'Credibility and trust are highly valuable tokens of trade when dealing in the political environments or organizations … When you are perceived as being believable and honest (credible), and reliable and responsible (trustworthy), managers will rely upon you more than upon others. Peers and members of your team will seek your advice and input and often take you into their confidence.' (Holden, 2003)

On the matter of credibility, Gerry Spence, one of America's greatest trial lawyers, who has never lost a single case in 40 years, once said to a jury, 'One can stand as the greatest orator the world has known, possess the quickest mind, employ the cleverest psychology, and have mastered all the technical devices of argument, but if one is not credible one might just as well preach to the pelicans.' (Mills, 1999)

18.3.2   Issues in collaborative work

In any collaborative work, one must recognize who the players are and what their functions and contribution are. David Strauss reminds us that there are four types of players (Strauss, 2002). There are those who can make decisions, but there are also those with the power to block decisions. Then there are those whose business units will be affected by the decisions, and those who hold relevant information for the decision-making process. The important job for the process leader is to recognize the dynamics of these four groups of people, and to steer the process towards the common goal.

A skilful process leader works like a diplomat, negotiating, prompting and navigating through the four types of stakeholders to achieve the goal; he or she manoeuvres between obstacles and turns them into productive and informative data and consensus.

18.3.3   Process leader as a communicator

Jim Clemmer noted that what are often called 'communication problems' in many groups or organizations are really problems of process, system, or structure (Clemmer, 1999). If people fail to communicate, it is because the way they are organized does not allow them do it effectively. Organizations that do not communicate well internally have not developed and fostered a corporate culture and environment for effective communications. In today's fast-paced world, business people do not have the time and patience to decode messages that do not get right to the point. Simplicity, clarity and accuracy are the critical success factors in communication.

18.3.4   Listening as a key element in communication

A golden rule is to learn to listen well before you communicate. Listening is a complex skill and must be developed to be an effective communicator. When you listen, you listen beyond words. It involves not only understanding the content of the conversation, but also the relational message embedded in words and gestures (Held, 1999). Emotional intelligence plays an important part in business and social communications. Learn to listen and respond in empathy. Reinforce the response by positive questioning or affirmation, confirmation of understanding and supportive body language. Consider the routine in a surgeon's operating room as illustrated below:

'Think ... of a surgeon calling for a tool: the surgeon states the name of the instrument required. Then the OR assistant retrieves the tool, repeats the name of the tool, and places it into the hand of the surgeon with a particular amount of force and in a specific manner to ensure both parties share full understanding and agreement.' (Prentice, 2002)

Facility managers rarely do this. A good practice in any collaborative work is to confirm the message received, to ensure the original meaning and intentions are understood. Repeating or restating the message in one's own words can accomplish this. Data must be confirmed for accuracy.

18.3.5   Dealing with difficult players

Teams and people are not perfect. This is apparent when competent professionals come together to form a building team. However, everybody wants to work with a perfect team. The idea of pulling a team of experts or stakeholders together does not necessarily mean that all these people will work together without disagreements or arguments. Harvey Robbins and Michael Finley remarked in their book *Why Teams Don't Work*, 'Ideal teams are comprised of perfect people, whose egos and individuality have been subsumed into the greater goal of the team. Real teams are made up of living, breathing, and very imperfect people.' (Robbins and Finley, 1997)

Personality plays a part in creating difficulties in team-work. Understanding the personality types of the team members helps to promote a better understanding of why certain people behave in a certain way. However, the problem typically lies in the quality of listening skills. When a person is tuned out or not listening, or worse, refuses to listen, then communication cannot occur. The mind is switched on 'auto pilot'. (McKenna and Maister, 2002)

Concerning verbal communication, studies by the Harvard negotiation project (Stone, Patton and Heen, 1999) have revealed that there is a basic structure in difficult conversations. It can be categorized into three types:

- The 'What Happened?' conversation – this is the *who-is-to-blame* syndrome and is very destructive to relationship building between team members.
- The feelings conversation – emotions like anxiety, hurt or anger take over the conversation; sometimes such emotions are hidden, causing more misunderstanding and disruptions.

● The identity conversation – this is a self-directed conversation where one creates guilt, incompetence, and doubts within oneself, losing the reality of the situation.

18.4   Interpersonal skills in process leadership

A successful manager requires other skills, in addition to technical abilities, related to administration, communication, interpersonal relationship, leadership, motivation, organization, thinking and self-management (Davis et al., 1992). Effective communication skills form the foundation for successful management. It is important to realize that communication skills interact with other skills and serve as the translating mode for thoughts, ideas and action. In order to be successful, it is essential to develop all these skills and to learn how they are interoperative with one another. This is of particular importance for the process leader.

Robert Bolton, author of *People Skills*, claims the reason for an 80 per cent failure rate of projects at work is due to people's inability to relate to other people (Bolton, 1986). This may be due in part to the organization's negligence in promoting healthy emotional hygiene, and in part to a lack of proper and effective communication.

18.4.1   The importance of paralanguage

Former Press Secretary to HM The Queen, and now Visiting Professor of Personal and Corporate Communications at the University of Strathclyde, Michael Shea, suggested, 'The visual impact of the human communication process is greatly underestimated.... In many communications between two private individuals, or between a speaker and his or her audience, well over 50 per cent of the message is delivered by the image and body language of the speaker, around 40 per cent by the character, timbre and strength of the voice; that is, the elements of sound other than the words, or, in other words, their paralanguage; and amazingly, only 10 per cent or less by the words themselves.' (Shea, 1998)

A study published in 1967 by UCLA's Albert Mehrabian measured the impact of non-verbal body talk which Shea referred to as paralanguage, and found that it accounts for 93 per cent, and the message in words spoken only the remaining 7 per cent. However, further studies indicate that in many cases, when the message is critical and important, it accounts for 53 per cent, while the non-verbal body talk represents 47 per cent. However, Harry Mills, in his book *Artful Persuasion*, notes that when the message is 'information heavy', such as in a face-to-face sale or negotiation, then the content of the message becomes much more important (Mills, 1999). He warns: 'One of the biggest mistakes you can make in observing body language is to make judgements on the basis of a solitary gesture or to ignore the context. Gestures come in clusters and should always be interpreted this way.'

18.5   The emotional intelligence (EQ) of the
      process leader

Jack Welch, referred to by some as the world's greatest turnaround king, acknowledged that the most painful experience in his transformation process at GE was to handle 'the

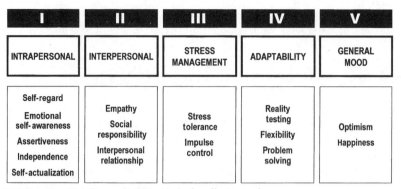

Figure 18.3 Emotional intelligence building blocks for effective performance.

influence of emotions' on the organization (*Financial Times*, Dec. 13, 1999). Recent research has shown that emotional intelligence or wellness can now be measured in terms of an 'emotional quotient' (EQ), similar to the intelligence quotient (IQ) that has been in use for years.

In the past, researchers have difficulty explaining why some people with superior IQ seem to fail in life, while others with lower IQs seem to succeed. Recent research has revealed that people with greater EQ are better able to achieve success. Dr Reuven Bar-On, the former Director at the Institute of Applied Emotional Intelligence in Denmark, spent over 20 years studying the relationship between success and emotion in individuals. He coined the term 'EQ' in 1985. In 1990, Dr Peter Salovey of Yale University and Dr John D. Mayer of the University of New Hampshire developed a framework to study how emotions facilitate adaptive cognitive and behavioural functioning, which they called 'emotional intelligence'. Daniel Goleman later popularized EQ with his 1995 bestseller '*Emotional Intelligence.*'

Bar-On concluded that emotionally intelligent people possess positive self-regard, can actualize their potential capacities and lead fairly happy lives. They can understand the way others feel. They can make and maintain mutually satisfying and responsible interpersonal relationships, without becoming dependent on others. They are generally optimistic, flexible, realistic, and successful in solving problems and coping with stress, without losing control. He created a behavioural measurement called the emotional quotient inventory (Bar-On EQ-I), published in 1997 as the world's leading scientific instrument for EQ, which has been tested on well over 100 000 people worldwide.

The EQ competencies for effective performance in the workplace, according to the Bar-On EQ-I methodology, are based on five building blocks: intrapersonal, interpersonal, stress management, adaptability, and general mood. He further sub-divides these into fifteen emotional skill sets: self-regard, emotional self-awareness, assertiveness, independence, self-actualization, empathy, social responsibility, interpersonal relationship, stress tolerance, impulse control, reality testing, flexibility, problem solving, optimism and happiness.

18.6   Key EQ competencies for the process leader

In 2003, the author conducted a survey to determine the top five EQ competencies required for facility managers. Over 250 facility professionals responded, and the results

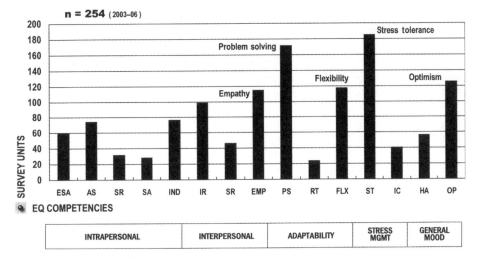

Figure 18.4 EQ competencies survey for facilities managers (2003).

are: empathy, problem-solving, flexibility, stress tolerance and optimism. It is interesting to note that further sorting by gender yielded the same results. Each of these traits is discussed below.

18.6.1

Empathy – empathy is the ability to understand another person's feelings and situation. A big part of this is built upon good listening skills – empathic listening. Most people are always too busy rushing from one place to another, or from one thought to another, that they never slow down enough to really listen to the conversation or message. Stephen Covey in *The 7 Habits of Highly Effective People* said, 'empathic listening gets inside another person's frame of reference'. (Covey, 1989) In empathy, there is a need to switch to the other person's paradigm.

In collaborative work, where the emphasis is on performance and speed, people are not often able to spend enough time and attention to ensure cohesive functioning between project members. The essential ingredient to a successful team is empathy. Empathy is like 'glue' that binds the emotional experiences of the professional members together and breaks down egocentric behaviours.

18.6.2

Problem solving – many process leaders and professionals score high in this area. The facility management business is a problem-solving business; we are surrounded by new situations and challenges every day. Problem solving involves judgement and decision-making. There is a dynamic relationship between cognitive activities, tacit knowledge and emotional activities in an individual during the problem-solving process (Bar-On and Parker, 2000). This is particularly important within a group of knowledge workers like architects and engineers.

In the context of EQ, problem-solving skills go beyond everyday work situations. Problem solving must engage superior interpersonal and empathetic skills. It involves solving personal and interpersonal problems (Ciarrochi, Forgas, and Mayer, 2001). To maintain a high standard in the six-phase integrative building performance process, problem solving must be realistic: the solution must meet the needs. The value-benefits of the solution will generally have long-term implications for the business.

18.6.3

Flexibility – Mel Silberman in *PeopleSmart* is convinced that, to improve relationships, you must be able to 'shift gears', meaning that a change in one's behaviour is often the catalyst for a change in the other person's behaviour (Silberman and Hansburg, 2000). Being flexible is being able to shift gears – adjusting emotions, ideas, thoughts and behaviour to ever-changing situations and environments. In construction, job site conditions produce daily problems, not unlike the business interruptions and abnormalities encountered by business leaders. Flexibility is therefore a key competency for building professionals.

18.6.4

Stress Management – CNN reported in March 2001: 'Many US workers may be working too hard, leading to more mistakes on the job, neglected personal relationships and higher healthcare costs.' The New York-based non-profit Family and Work Institute conducted a study showing that 46 per cent of the respondents said they felt overworked in one way or another. Ellen Galinsky, institute president and co-author of the study, suggested that this 'study is a clarion call for all of us – companies and individuals – to look at how we're working'.

People with the skill set for stress management are able to find their way out of adverse and stressful situations without being mentally drained and distressed. People with poor or low tolerance for stress will lose control, resulting in poor judgements and decisions. They can become confused, disoriented, out of focus, and burned out. This is particularly important in building delivery, where each professional is a subject matter expert and all are eager to offer their version of a solution. Process leaders that can coach team members to manage stress and control behavioural impulses will increase collaboration and achieve results.

18.6.5

Optimism – according to the Collins *Cobuild English Dictionary*, optimism is the feeling of being hopeful about the future or about the success of something in particular. The two key words are 'hopeful' and 'success'. It seems that people who possess these two words in their emotional vocabulary tend to have a more positive outlook on life and a better expression in their emotions. Dr Steven Stein calls it the 'ability to look at the brighter side of life and to maintain a positive attitude, even in the face of adversity' (Stein and Book, 2000).

Everything people experience, whether pleasant or not, becomes information in the mind. If one is trained to control this information, and can control one's inner life, one can achieve happiness. According to Dr Mihaly Csikszentmihalyi, professor and former chairman of the Department of Psychology at the University of Chicago, happiness is an inner condition that must be prepared for, cultivated, and defended privately by each person (Csikszentmihalyi, 1990). The state of happiness can have a profound effect on how one thinks and how one works. Happiness and optimism are therefore closely related.

In the workplace, optimistic people can generally cope better than others with more stress, changes, and uncertainties. As optimism fosters happiness, it is the job of the process leader to create a team culture that embraces a positive attitude and outlook on life in order to attain effective performance. These are the people who will generally make things happen, even in adverse conditions.

18.7   Conclusions

The building delivery process is a complex process. It involves many parties and many professionals, with an increasing number of specializations. The success of a building project does not depend on how many professionals are involved, but on how well these people relate to one another, and how well they work together towards a shared vision of an integrated product.

The quality of the solution depends on the quality of instructions and specifications given and received. This is the paramount responsibility of the process leader. The process leader ensures the accurate transfer of ideas, concepts and data that affect decision-making. The process is about orchestrating the interrelationship of team members. The ability to motivate and stimulate each professional to perform optimally depends on the skills and competencies of the process leader. This process refers to more than technical performance, to include the emotional and social performance of individuals and teams. The human element is often forgotten in technically intensive work, but it is of utmost importance in the success of reaching the solution and, in this case, realizing the integrated building delivery process.

References

Bar-On, R. and Parker, J.D.A. (eds) (2000). *The Handbook of Emotional Intelligence*. Jossey-Bass Inc.

Bennis, W. (1999). *Managing People is like Herding Cats*. Executive Excellence Publishing, Provo, UT.

Bolton, R. (1986). *People Skills*. Touchstone Edition.

Ciarrochi, J., Forgas, J.P. and Mayer, J.D. (2001). *Emotional Intelligence in Everyday Life – A Scientific Inquiry*. Psychology Press.

Clemmer, J. (1999). *Growing the distance*. TCG Press, Kitchener.

Covey, S. (1989). *The 7 Habits of Highly Effective People*. Simon & Schuster.

Csikszentmihalyi, M. (1990). *Flow – The Psychology of Optimal Experience*. Harper Perennial.

Damasio, A.R. (1994). *Descartes' Error – Emotion, Reason, and the Human Brain*. Avon Books.

Davis, B.L., Skube, C.J., Hellervik, L.W., Gebelein, S.H. and Sheard, J.L. (1992). *Successful Manager's Handbook*. Personnel Decisions, Inc.: Minneapolis, MN.

Harris, S. (2003). www.brainyquote.com.

Held, V. (1998). *How Not to Take it Personally*. VNH Communications: Toronto.

Holden, M. (2003). *The Use and Abuse of Office Politics – how to survive and thrive in the corporate jungle*. Allen & Unwin: Crows Nest, NSW, Australia.

McKenna, P.J. and Maister, D.H. (2002). *First Among Equals*. The Free Press.

Mills, H. (1999). *Artful Persuasion – how to command attention, change minds and influence people*. MG Press: Sumner Park, Australia.

Prentice, S. (2002). *Cool Time and the Two-pound Bucket*. Stoddart.

Robbins, H. and Finley, M. (1997). *Why Teams Don't Work*. Orion Business.

Shea, M. (1998). *The Primacy Effect – the ultimate guide to effective personal communications*. Orion House.

Silberman, M. and Hansburg, F. (2000). *PeopleSmart – developing your interpersonal intelligence*. Berrett-Koehler Publishers, Inc.

Stein, S.J. and Book, H.E. (2000). *The EQ Edge*. Stoddart.

Stone, D., Patton, B. and Heen, S. (1999). *Difficult Conversations – how to discuss what matters most*. Viking Penguin.

Strauss, D. (2002). *How to Make Collaboration Work*. Berrett-Koehler Publishers, Inc.

PART FOUR

Epilogue

Looking to the future

Jacqueline C. Vischer

19.1 Assessing building performance

This book is dedicated to outlining, describing and illustrating an innovative approach to the way buildings are planned, designed, built, and managed after occupancy. The approach is innovative because it proposes a logical, sequential and cyclical process of decision-making throughout the process. It is oriented to ensuring the production of high quality buildings, and it places human needs and uses of space at the centre of this process. Typical approaches to producing and operating buildings do none of these things.

Several authors in this volume, including Preiser, Holtz, Bordass and Leaman, and Ornstein et al., characterize the typical building delivery and occupancy process as:

- *disjointed*, as each professional specialist or contractor makes decisions independently;
- *cost-driven*, as decisions pertain almost uniquely to construction rather than operating costs, and therefore to a short-term perspective rather than the lifetime of the building;
- *time-limited*, as taking time to incorporate additional information, to consider alternatives and to review possibilities, is disadvantageous in the cost-driven process;
- *conflict-ridden*, as the architect's priorities – supply-side concerns to make decisions quickly, reduce risk and encourage decision-making with graphics and visual aids – differ from the contractor's priorities to use established and well-known methods and materials to reduce risk and save time. These again may not coincide with the client's or users' priorities and demand-side concerns to keep options open as long as possible, to make sure all information is gathered, and maximize flexibility (Blyth and Worthington, 2001).
- *ignorant*, both of human needs and of the human use of buildings.

The result is often that commercial buildings are inflexible, costly to operate, uncomfortable, not ecologically responsible and ultimately require significant reinvestment to 'make them work'. This process, which has been characterized as dysfunctional by many writers, is a result of the way the real estate industry is structured in our society. Several authors in this book refer, for example, to the fragmentation of the building professions and the emergence of new specialties. Traditionally, the architect or master-builder exercised control over the entire building process and could, by implication, ensure a quality result.

Now the architect is one of a series of specialists involved in a complex and expensive project, over which no single participant exercises control, including the relatively 'new' professions of project management, facilities management, interior design and space planning. Moreover, these specialists are not trained to be process leaders.

The way the industry operates is also influenced by the way financing institutions operate. Capital funds are loaned under conditions that favour speed over thoroughness, are limited to the short-term time-frame of building delivery without reference to the long-term considerations of building operation, and discourage innovation and risk at all levels. Little data are yet available on the financial and social costs to society of building and occupying buildings that do not work for users and that ignore issues of quality and sustainability, although some are raised by Roaf in Chapter 9.

The building performance evaluation (BPE) approach presented by Preiser and Vischer in Chapter 1 and described by Preiser and Schramm in Chapter 2 proposes a restructuring of building delivery and operations. BPE ensures that feedback from stakeholders – not limited to the building's occupants, but also facilities management, client/owners and lending institutions, as well as professionals in the building industry – informs each stage in the process in a systematic way. It is important to note, as the chapters in this book demonstrate, that none of these phases, feedback loops and information tools is new. What is new is putting them together and presenting them as a whole, and as a holistic approach to restructuring the way buildings are built and managed. Books and articles on related topics, such as improving ventilation and indoor air quality in office buildings, increasing sensitivity to users needs, linking office design with productivity, and making sustainability issues a priority, are proliferating. The BPE approach is a comprehensive and overarching methodology of ensuring that these, as well as other issues related to building quality, are not just considered separately and occasionally at the whim of the client or the project manager, but actually drive the process.

19.2   Phases of building performance evaluation

In Chapter 3, Schramm describes a strategic planning process, and provides an example in which clients drew on professional expertise from architects and planners before making the first, critical decisions about their new building. The data gathered at later phases in the project indicate that this was time well invested. A wrong or bad decision at an early stage creates problems and costs throughout the rest of the process. Schramm's examples underline not just the obvious importance, but also the tangible value, of investing in strategic planning, Phase 1 of BPE.

Chapters 4, 5 and 6 describe and illustrate subsequent phases of BPE (briefing/programming and design review, as well as construction and commissioning). They provide incontrovertible examples of failures that occur when these phases are not performed thoroughly and well. Marmot et al. indicate how many costly mistakes can be avoided in later stages of a building project by thorough and systematic analysis of users' needs, resolution of potential conflicts, establishing priorities, and enumerating performance criteria during Phase 2, programming/briefing. Vischer describes an example of how involving users in design review during Phase 3, design, in a managed participatory process, resulted in a high quality building, provided innovative workspace to users who fully accepted and used it, and cost no more than a conventional office building project. Holtz elaborates a commissioning methodology for Phase 4, construction, that is designed to ensure that at each phase the client/owner gets the building that was expected and anticipated. Again, if implemented correctly, feedback from testing and inspections, from referring back to programming criteria, and from checklists and databases, is increasingly sought as a building project nears completion and moves into the occupancy phase.

Applying BPE means that each phase consists of one or more feedback loops, such that relevant information from stakeholders is sought out and applied to important building decisions. In none of the examples offered did the use of feedback procedures increase project costs. On the contrary, they were less than might have been expected. And why? Because rational decision-making based on the right information was built into each of the examples described. Once a building is built and occupied, other tools are available to optimize building operations and continue to ensure a quality experience for users (Vischer, 1989). The principal tool for Phase 5, occupancy, is post-occupancy evaluation (POE). In Chapter 7, Bordass and Leaman show how POE studies have evolved in recent years, and describe repeated efforts in the UK to standardize them as part of the building delivery process. Many of the chapters in Part 3 provide examples of POE feedback loops applied in a wide variety of contexts, illustrating the many ways in which user feedback on the buildings they occupy has a number of important uses.

Among other things, POE is a tool for use by facilities managers during the last and longest phase of the BPE process that lasts throughout the years of occupancy, renovations, and reuse of building interiors, all the way to complete recycling of the building in some other form. Then outlines in Chapter 8 how the role and responsibilities of facilities managers can be aided and informed by measuring the technical performance of building systems systematically, as well as incorporating feedback from users. The author explains the traditional tendency in business to see buildings as a cost rather than a resource. However, with more information available, and feedback loops in place to ensure input from users, owners and building system operations, a changeover in the business perspective to seeing buildings as a vital business resource seems to be under way.

19.3   BPE in a diversity of cultural contexts

Several of the case studies in Part 3 provide tools, methods and examples from their respective cultural contexts of how to advance and encourage the BPE perspective on office space acquisition and operation. For example, in Chapters 11, 12, 13 and 15 accounts from Germany, Brazil, Israel and the Netherlands, respectively, describe how systematic feedback from building users was used to provide constructive and needed support to owners and managers. In Germany, Walden illustrates a direct link between office environmental design and employee productivity, and offers a perspective not only on present, but also on future user needs in offices. In Brazil, Ornstein et al. have systematically collected information about use and occupancy of office buildings which, it is anticipated, will influence the building supply industry to change its priorities and address issues of building quality as well as building appearance. In Israel, Windsor recounts how feedback from building users helped to facilitate employees' acceptance of a new work environment in a difficult situation of innovation and resistance to change, so as to ensure optimal task performance. And, in the Netherlands, Mallory-Hill et al. have used a systematic approach to user feedback to facilitate the introduction of workspace innovations such as non-territorial office design and 'intelligent' daylighting. Each of these instances shows how appropriate user feedback, applied at the right time, in several different national contexts, could or did result in financial savings for building owners, developers and managers by increasing quality for users.

Chapters 10 and 14 describe tools and techniques for gathering building performance feedback in a variety of situations and contexts. This feedback can include, but is not limited to, cost data. However, finding ways to meet users' needs, improve quality and incorporate feedback results in streamlining operations at numerous levels, with an important effect on building operating costs. In Chapter 10, Szigeti et al. describe an approach to gathering and analysing data on a rich array of building performance indicators, to help owners, managers and designers make rational building operating and asset disposition decisions that are oriented towards balancing costs and quality. And in Chapter 14, Kato et al. indicate how innovative methods of gathering data on how people use space in the buildings they work in, results in new information that can affect organizational decision-making. Such feedback from users advises facility managers on building system performance, and also shows business managers how their employees use space, thus enabling more functional and more efficient space planning inside office buildings.

Looking to the future, Preiser links BPE and universal design in Chapter 16; and in Chapter 17, Zimring et al. outline a promising public sector initiative to try to make BPE – specifically, building evaluation – a rational and routine way for public buildings to demonstrate an approach to improving building quality. Finally, and importantly, in Chapter 18, Lam outlines ways of training people to lead and facilitate the kinds of group communication processes that make BPE possible.

Most of these chapters provide examples of BPE phases and feedback loops in an international context, to show how a rational, user-feedback-based approach to buildings can have merit and meaning in many parts of the world, and for many different types of projects. It cannot then be charged that BPE only works in one culture or context; on the contrary, BPE offers a broad and adaptable framework for professionals affiliated with the building industry at all levels to find ways of implementing a user-oriented, cost-effective and high quality approach to producing all types of buildings. In effect, BPE represents an optimistic and common-sense option for the future, the implementation of which can and will lead to better quality buildings and a better organized and functioning building industry.

19.4   Quality and cost

The building industry as it currently operates in most countries is almost entirely cost-driven. Cost calculations are applied to every stage of construction, and time is money. Thus, on some projects, such as office buildings, which are built to pre-set standards to meet zoning requirements and respond to market demand, programming is not considered a necessity; whereas on others with complex functions, such as hospitals, programming is confined to technical performance requirements and systems specifications. Information-gathering initiatives, such as programming, post-occupancy studies, and using sustainable materials and processes, take extra time if they are 'added on' incrementally to a pre-existing and linear building delivery process. Most building project cost-estimating is based on a series of sequential steps related to estimates of the time needed to complete them. Taking time to acquire more or better information, seek feedback and otherwise inform decisions, adds to the length of time needed until the building is built and occupied and starts generating revenues, and thus to the cost of the financing needed for the project.

After the building is occupied, cost considerations continue to drive building operating decisions. A major part of the facility manager's task is to drive down energy costs, 'churn' costs incurred from reconfiguring workspace in order to move people around, maintenance and repair costs, and to balance operating plus capital costs with possible income generated by leasing out space. Facility managers benchmark their costs against those of other companies and buildings, using pre-set performance indicators that provide an assessment of whether or not the building is efficiently run. This defines 'quality' according to the cost-driven perspective.

However, these cost evaluation frameworks often fail to address other definitions of quality, or even other costs. Unanticipated costs arise as a result of the separation between construction financing and operating budgets that is built into the traditional approach. Thus, all repair and renovation work that must be done after a building is occupied – to correct performance problems, to increase occupants' comfort, and to accommodate changing uses and needs – is not calculated as part of the cost of delivering the building; these are considered operating costs. Moreover, if certain of the building systems do not function as expected, the cost of time lost by occupants who cannot use a space or a piece of equipment is rarely added to the cost of repair or replacement. This 'downtime' signifies the negative economic value of disruption of work processes caused by the facility infrastructure not functioning properly.

Downtime can be caused by calamities such as fire, as well as by information technology failures, or even a badly planned move; it is related to the concept of flexibility. Flexibility indicates the ease with which elements of the facility infrastructure can be changed to support organizational changes (Wagenberg, 2004). How efficiently, in terms of cost and time, can the organization make change, and employees be adequately supported in newly reconfigured space?

As most modern buildings built in traditional ways lack the ability to adapt easily and cheaply to changing functional demands, lack of flexibility adds to operating costs. But the decisions that would have made a building 'flexible' are made while a building is being planned, designed and financed, and not after it is occupied. Decisions that would have increased building quality, for example, by providing flexibility, are not taken at the appropriate time, either through ignorance, or because a cheaper option is chosen. Flexibility is that building attribute that enables the organization's activities to be supported into the future, even as technology and business processes are changing. Flexibility thus has value to the occupying organization, and lack of it increases costs.

Flexibility is an important ingredient of building quality – that elusive and much-touted abstraction that refers to how well users' and clients' needs are met. If the building industry does not always build the cheapest building with the cheapest materials, it is because clients and tenants attach value to quality. This goes beyond the value for an individual user. The meaning of the concept of 'quality' for facilities managers is related to the needs and requirements of their clients and of the building users that are their customers. Through being attached to and involved with the organization, facility managers usually come to know what their customers need, and they translate this into quality criteria, such as service level agreements which, in an ideal situation, they can monitor.

Several years ago (1989), this author wrote, 'Environmental quality is that combination of environmental elements that interact with users of the environment to enable that environment to be the best possible one for the activities that go on in it.' (Vischer, 1989)

For building developers, quality in buildings can mean concessions to a higher level of comfort for occupants that clients will be willing to pay for. According to the BPE approach, however, quality is not an extra that clients have to buy, but rather a basic responsibility of building providers and managers towards the users of buildings. Replacing time and money with quality at the centre of the building construction and operation process would do away with the short-term, cost-driven, piecemeal and linear approach to delivering and managing buildings, and therefore with the added cost associated with managing buildings resulting from this process. It would also do away with the notion that taking time to acquire information and feedback to inform decisions is not a cost-effective process.

The BPE approach is cyclical rather than linear. The purpose of the feedback loops built into the six BPE phases is to provide opportunities and mechanisms for gathering information on which decisions about quality can be based. Quality is composed partly of generic facts about human comfort generally – and human comfort in buildings – and partly of information that is specific to a situation or project. Hence Chapters 1 and 2 draw attention to the value of gathering generic information from data banks, such as design guidelines, codes and standards, research results and the like, as well as applying user surveys, focus groups, mapping and other techniques for data on users in specific situations. The BPE approach ensures that opportunities not just to collect, but also to apply feedback from users are built into the process on a systematic basis.

How, then might one expect a building industry to change from what we have now, to one focused on quality? All the phases of strategic planning, programming, design, construction, occupancy and adaptive reuse/recycling will be carried out routinely and legit-imately on every project, and not just in some situations with special requirements. Financial lenders will calculate their loans and incentives for new building projects, not just according to short-term returns, but by taking quality criteria for long-term building operation into consideration. The time invested in ensuring all the feedback loops of BPE phases are implemented will be more than compensated by the reduced time spent later on in the life of the building correcting problems, repairing elements that do not work, and adapting to new uses. Users will feel empowered instead of imposed upon: one of the spin-off effects of consulting them about their needs will be better informed users, and people capable of and interested in participating in decisions about their environment. This will translate into fewer complaints and less service calls, and a constructive partnership between users and facilities managers.

19.5   Vision of the future

As the chapters in this book make clear, the component parts of BPE are already in existence and proven in many parts of the world. In order to see a complete implementation of the BPE approach, the process of linking them up must be set in motion by competent and experienced professionals. They must ensure that decision-makers are involved in each of the phases, and carry through on applying user feedback to building design, construction and operation. The qualities of such leaders are well set out in Chapter 18. Training for leading the BPE process is available to all building professionals, including architects, project managers, interior designers and facility managers. However, it is the latter group that may have the most to offer in terms of knowledge that could be applied to the entire process. With their perspective on the human and technical aspects of building operation, facility managers could inform design and construction decision-making. And with the close contact with occupants that results from managing the building, facility managers are well placed to understand and anticipate changing user needs.

For the future, it will be necessary to demonstrate the benefits and advantages of anchoring BPE in a real-world context, such that it becomes the approach of choice for the building industry. Case studies that measure the results of a different and better building delivery process (planning, briefing, design review, commissioning), that indicate the rewards of applying universal design principles, and that continue to measure and evaluate the effects of building design decisions on use and users, are needed to increase pressure on present industry practices. As the structure and content of such case studies are standardized, accessible databases can be built up, similar to the public sector initiative described in Chapter 17, and compelling results made available to all stakeholders.

In view of the initiatives described in this book, already established in numerous countries, and the number of other books and projects they in turn refer to, we can only speculate for how much longer the commercial building industry will continue in its dysfunctional ways. Small gains are being made all over the world, and a growing number of people outside the traditional closed circle of financier-developer-designer-builder are becoming involved both in supplying buildings, and in examining details of their operation over their lifetimes. These include researchers, process facilitators, non-traditional architects, interior architects and managers, as well as members of the new group known as 'work environment specialists'.

Some of these perspectives are technical, some are cost-oriented, and some are humanistic – but all point in the same direction: change. The authors and researchers who have contributed to this volume, and the myriads of people – students, technicians, building occupants – who stand behind them, are pioneers in this new and important social change. Together and separately, in countries all over the world, these people and others like them have dedicated major parts of their careers to efforts to increase the quality of the built environment – not just once, but on a systematic basis, and not just for building users, but also for owners, builders, project and facility managers, and members of the design disciplines. It is to be hoped that the publication of this book, and global recognition of the work of these authors and others like them, will increase their faith and commitment that change is possible. In time, ever-increasing numbers of clients, building professionals and the lay public will begin to see the value and importance of the BPE perspective.

Acknowledgements

Thanks to Andreas van Wagenberg for his contribution to this chapter.

References

Blythe, A. and Worthington, J. (2001). *Managing the Brief for Better Design*. Spon Press.
Vischer, J.C. (1989). *Environmental Quality in Offices*. Van Nostrand Reinhold.
Wagenberg, A.F. van (2004). Facility management: an introduction to the Dutch perspective and practice. www.fm.chalmers.se/kursl-l/DvW.pdf

Appendix: Measuring instruments for building performance evaluation

Several of the contributors to this volume have supplemented their chapters with measuring instruments and tools that they have developed and/or used for measuring different phases of building performance evaluation.

In the following pages, instruments used in the Netherlands, Germany, Canada, Brazil and USA are presented, translated into English where appropriate. Several are in the form of survey questionnaires, to be administered to building users, although the areas of investigation targeted by the different surveys are not always the same. One looks at all aspects of the indoor environment that are likely to impact user satisfaction. One measures user behaviour in office buildings. One evaluates the impact of innovative design on building users. One is limited to assessing users' comfort relative to interior environmental conditions. Others are in the form of information sources, such as a summary of design principles, and useful web sites in which additional measurement tools and approaches can be found.

Readers are encouraged to test these tools in their own settings, and to compare them with methods developed and published elsewhere. The field of BPE is in constant evolution and changing fast, thus more methodological sophistication and ways of capturing and analysing data can be, and are, added at any time. The editors of this volume would be pleased and interested to receive feedback from readers on the usefulness and adaptability of these tools, as well as on other tools and skills developed by researchers and professionals who are not contributors to this book (see author contact information at the beginning of this book).

In this Appendix, the tools that have been included are in the following order. First (A1), a checklist of useful information necessary for effective BPE is provided, along with a list of key questions for building owners/managers. Then (A2), several user survey questionnaires are listed, starting with a generic BPE questionnaire for building users (referred to in Chapter 1). This is followed by a survey tool (A3) developed in Canada, and described in Chapter 2; a questionnaire (A4) developed in Brazil and described in Chapter 12; and two different user questionnaire surveys (A5, A6) developed and used in the Netherlands, as described in Chapter 15. One provides a general evaluative framework for user comments and feedback; the other is specifically oriented to the impact of innovative workspace design on users. Next (A7), several measurement tools in use in Germany, as described in Chapter 11, are included. These include the mapping tool for office environments, the system to judge office quality (using trained observers), and a detailed user survey questionnaire. To conclude, generic principles of universal design are listed (A8) (as described in Chapter 16); and, finally, a list of information sources for building commissioning (A9), to complement Chapter 6. In addition, two chapters (5 and 8) refer to a web site with up-to-date information on new and useful approaches to briefing (programming) and post-occupancy evaluation.

We hope that, by making tools and information easily available, more professionals and researchers involved in the building industry will test and try them out, thereby making feedback loops a fundamental part of the industry's approach to planning, designing and managing buildings.

A.1 Checklist of useful documents for building performance evaluation (BPE)[1]

Client-related information

1. Client mission statement, organizational chart, and staffing.
2. Initial programme for the building.
3. As-built floor plans (may require updating).
4. Space assignments and schedules.
5. Building-related accident reports.
6. Records of theft, vandalism, and security problems.
7. Maintenance/repair records.
8. Energy audits or review comments from heating/cooling plant manager.
9. Any other feedback concerning the building, which may be on record.

Building-related information

1. Identification of selected recent, similar facilities.
2. Review of programmes and other pertinent information on the building type being evaluated.
3. Identification and assessment of state-of-the-art literature (e.g. technical manuals and design guides).

Building performance evaluation (BPE) questions

We would like to know how well your building performs for all those who occupy it. Successes and failures (if any) are considered insofar as they affect occupant health, safety, efficient functioning, and psychological well-being. Your answers will help improve the design of future, similar buildings.

Below please identify successes and failures in the building by responding to the following broad information categories, and by referring to documented evidence or specific building areas whenever possible.

1. Adequacy of overall design concept.
2. Adequacy of site design.
3. Adequacy of health/safety provisions.
4. Adequacy of security provisions.
5. Attractiveness of exterior appearance.
6. Attractiveness of interior appearance.
7. Adequacy of activity spaces.
8. Adequacy of spatial relationships.
9. Adequacy of circulation area, e.g. lobby, hallways, stairs, etc.
10. Adequacy of heating/cooling and ventilation.
11. Adequacy of lighting and acoustics.
12. Adequacy of plumbing/electrical.
13. Adequacy of surface materials, e.g. floors, walls, ceilings, etc.
14. Underused of overcrowded spaces.
15. Compliance with ADA handicapped accessibility requirements.
16. Other, please specify (e.g. needed facilities currently lacking).

[1] International Building Performance Evaluation (IBPE) Consortium (1995). University of Cincinnati, OH, USA

A.2   Occupant survey

We wish to conduct a performance evaluation of your building. The purpose of this evaluation is to assess how well the building performs for those who occupy it in terms of health, safety, security, functionality, and psychological comfort. The benefits of building performance evaluation include: Identification of good and bad performance aspects of the building, better building utilization, feedback on how to improve future, similar buildings, or remodelling of your own building.

In the survey that follows, please respond only to those questions that are applicable to you. Indicate your answers by marking the appropriate blanks with an 'X'.

1. In an average work week, how many hours do you spend in the following types of spaces? (specify):

Space A: _____

Space B: _____

Space C: _____

Space D: _____

Space E: _____

Hours	A	B	C	D	E
0–5	[ ]	[ ]	[ ]	[ ]	[ ]
6–10	[ ]	[ ]	[ ]	[ ]	[ ]
11–15	[ ]	[ ]	[ ]	[ ]	[ ]
16–20	[ ]	[ ]	[ ]	[ ]	[ ]
21–25	[ ]	[ ]	[ ]	[ ]	[ ]
26–30	[ ]	[ ]	[ ]	[ ]	[ ]
31–35	[ ]	[ ]	[ ]	[ ]	[ ]
35–40	[ ]	[ ]	[ ]	[ ]	[ ]
40+	[ ]	[ ]	[ ]	[ ]	[ ]

Key for the following quality ratings: EX = Excellent quality
G = Good quality
F = Fair quality
P = Poor quality

2. Please rate the overall quality of the following areas in the building:

	EX	G	F	P
(a) Space category A	[ ]	[ ]	[ ]	[ ]
(b) Space category B	[ ]	[ ]	[ ]	[ ]
(c) Space category C	[ ]	[ ]	[ ]	[ ]
(d) Space category D	[ ]	[ ]	[ ]	[ ]
(e) Space category E	[ ]	[ ]	[ ]	[ ]
(f) Restrooms	[ ]	[ ]	[ ]	[ ]
(g) Storage	[ ]	[ ]	[ ]	[ ]
(h) Elevator(s)	[ ]	[ ]	[ ]	[ ]
(i) Stairs/corridors	[ ]	[ ]	[ ]	[ ]
(j) Parking	[ ]	[ ]	[ ]	[ ]
(k) Other, specify	[ ]	[ ]	[ ]	[ ]

3. Please rate the overall quality of Space category A in terms of the following:

	EX	G	F	P
(a) Adequacy of space	[ ]	[ ]	[ ]	[ ]
(b) Lighting	[ ]	[ ]	[ ]	[ ]
(c) Acoustics	[ ]	[ ]	[ ]	[ ]
(d) Temperature	[ ]	[ ]	[ ]	[ ]
(e) Odour	[ ]	[ ]	[ ]	[ ]
(f) Aesthetic appeal	[ ]	[ ]	[ ]	[ ]
(g) Security	[ ]	[ ]	[ ]	[ ]
(h) Flexibility of use	[ ]	[ ]	[ ]	[ ]
(i) Other, specify	[ ]	[ ]	[ ]	[ ]

4. Please rate the overall quality of Space category B in terms of the following:

	EX	G	F	P
(a) Adequacy of space	[ ]	[ ]	[ ]	[ ]
(b) Lighting	[ ]	[ ]	[ ]	[ ]
(c) Acoustics	[ ]	[ ]	[ ]	[ ]
(d) Temperature	[ ]	[ ]	[ ]	[ ]
(e) Odour	[ ]	[ ]	[ ]	[ ]
(f) Aesthetic appeal	[ ]	[ ]	[ ]	[ ]
(g) Security	[ ]	[ ]	[ ]	[ ]
(h) Flexibility of use	[ ]	[ ]	[ ]	[ ]
(i) Other, specify	[ ]	[ ]	[ ]	[ ]

5. Please rate the overall quality of Space category C in terms of the following:

	EX	G	F	P
(a) Adequacy of space	[ ]	[ ]	[ ]	[ ]
(b) Lighting	[ ]	[ ]	[ ]	[ ]
(c) Acoustics	[ ]	[ ]	[ ]	[ ]
(d) Temperature	[ ]	[ ]	[ ]	[ ]
(e) Odour	[ ]	[ ]	[ ]	[ ]
(f) Aesthetic appeal	[ ]	[ ]	[ ]	[ ]
(g) Security	[ ]	[ ]	[ ]	[ ]
(h) Flexibility of use	[ ]	[ ]	[ ]	[ ]
(i) Other, specify	[ ]	[ ]	[ ]	[ ]

6. Please rate the overall quality of Space category D in terms of the following:

	EX	G	F	P
(a) Adequacy of space	[ ]	[ ]	[ ]	[ ]
(b) Lighting	[ ]	[ ]	[ ]	[ ]
(c) Acoustics	[ ]	[ ]	[ ]	[ ]
(d) Temperature	[ ]	[ ]	[ ]	[ ]
(e) Odour	[ ]	[ ]	[ ]	[ ]
(f) Aesthetic appeal	[ ]	[ ]	[ ]	[ ]
(g) Security	[ ]	[ ]	[ ]	[ ]

(h) Flexibility of use   [ ]   [ ]   [ ]   [ ]
(i) Other, specify   [ ]   [ ]   [ ]   [ ]

7. Please rate the overall quality of Space Category E in terms of the following:

	EX	G	F	P
(a) Adequacy of space	[ ]	[ ]	[ ]	[ ]
(b) Lighting	[ ]	[ ]	[ ]	[ ]
(c) Acoustics	[ ]	[ ]	[ ]	[ ]
(d) Temperature	[ ]	[ ]	[ ]	[ ]
(e) Odour	[ ]	[ ]	[ ]	[ ]
(f) Aesthetic appeal	[ ]	[ ]	[ ]	[ ]
(g) Security	[ ]	[ ]	[ ]	[ ]
(h) Flexibility of use	[ ]	[ ]	[ ]	[ ]
(i) Other, specify	[ ]	[ ]	[ ]	[ ]

8. Please rate the <u>overall quality</u> of design in this building:

	EX	G	F	P
(a) Aesthetic quality of exterior	[ ]	[ ]	[ ]	[ ]
(b) Aesthetic quality of interior	[ ]	[ ]	[ ]	[ ]
(c) Amount of space	[ ]	[ ]	[ ]	[ ]
(d) Environmental quality (lighting, acoustics, temperature, etc.)	[ ]	[ ]	[ ]	[ ]
(e) Proximity to views	[ ]	[ ]	[ ]	[ ]
(f) Adaptability to changing uses	[ ]	[ ]	[ ]	[ ]
(g) Security	[ ]	[ ]	[ ]	[ ]
(h) ADA compliance (handicapped accessibility)	[ ]	[ ]	[ ]	[ ]
(i) Maintenance	[ ]	[ ]	[ ]	[ ]
(j) Relationship of spaces/layout	[ ]	[ ]	[ ]	[ ]
(k) Quality of building materials				
(1) floors	[ ]	[ ]	[ ]	[ ]
(2) walls	[ ]	[ ]	[ ]	[ ]
(3) ceilings	[ ]	[ ]	[ ]	[ ]
(l) Other, specify	[ ]	[ ]	[ ]	[ ]

9. Please select and rank, in order of importance, facilities, which are needed but currently lacking in your building:

10. Please make any other suggestions you wish for physical or managerial improvements in your building:

11. Demographic information:
    (a) Your room #/building area: _____
    (b) Your position: _____
    (c) Your age: _____
    (d) Your sex: _____
    (e) # of years with the present organization: _____

A.3   BIU survey questionnaire

Date _____
Building number _____   *Floor* _____   *Workstation number* _____

You will find in the next few pages, a series of simple questions about your work environment. The questions are mostly in the form of a scale from 1 to 5, in which 1 means uncomfortable and 5 means comfortable. We will analyse your responses to provide a profile of the functional comfort of your work environment, compared to pre-existing norms. This information will be available to you, if you request it.

Please answer every question. Please do not discuss your replies with colleagues until after you have completed the survey. You may return completed questionnaires <u>within 24 hours</u> to the research team, or place them in the box for completed questionnaires located on your floor.

The *identification numbers on each survey form are to aid in data analysis. Please note that all individual responses are anonymous and will remain confidential.*

Thank-you for taking the time to complete this survey (we know you are often asked to fill out questionnaires).

Please assess the following aspects of your workspace on a scale of 1 to 5, where 1 means uncomfortable and 5 means comfortable and 3 means neutral. Please check one response to each question.

Temperature comfort:	1 UNCOMFORTABLE	2	3	4	5 COMFORTABLE	☐
How cold it gets:	1 TOO COLD	2	3	4	5 COMFORTABLE	☐
How warm it gets:	1 TOO WARM	2	3	4	5 COMFORTABLE	☐
Temperature shifts:	1 TOO FREQUENT	2	3	4	5 CONSTANT TEMPERATURE	☐
Ventilation comfort:	1 UNCOMFORTABLE	2	3	4	5 COMFORTABLE	☐
Air freshness:	1 STALE AIR	2	3	4	5 FRESH AIR	☐
Air movement:	1 STAGNANT AIR	2	3	4	5 GOOD CIRCULATION	☐
Noise distractions:	1 TOO DISTRACTING	2	3	4	5 COMFORTABLE	☐
Background noise levels:	1 TOO MUCH NOISE	2	3	4	5 COMFORTABLE	☐

Specific noises **(voices, equipment):**	1 TOO NOISY	2	3	4	5 COMFORTABLE	☐
Noise from the **ventilation systems:**	1 TOO NOISY	2	3	4	5 COMFORTABLE	☐
Noise from lights:	1 BUZZING LIGHTS	2	3	4	5 COMFORTABLE	☐
Noise from outside **the building:**	1 TOO NOISY	2	3	4	5 COMFORTABLE	☐
Furniture comfort in **your office/** **workstation:**	1 UNCOMFORTABLE	2	3	4	5 COMFORTABLE	☐
Size of your **office/workstation:**	1 UNCOMFORTABLE	2	3	4	5 COMFORTABLE	☐
Storage space in your **office/workstation:**	1 INADEQUATE	2	3	4	5 ADEQUATE	☐
Access to equipment:	1 UNCOMFORTABLE	2	3	4	5 COMFORTABLE	☐
Personal storage:	1 INADEQUATE	2	3	4	5 ADEQUATE	☐
Informal meeting- **spaces:**	1 INADEQUATE	2	3	4	5 ADEQUATE	☐
Space for collaborative **work with colleagues:**	1 INADEQUATE	2	3	4	5 ADEQUATE	☐
Space for meetings **with visitors:**	1 INADEQUATE	2	3	4	5 ADEQUATE	☐
Visual privacy:	1 UNCOMFORTABLE	2	3	4	5 COMFORTABLE	☐
Conversation **privacy:**	1 UNCOMFORTABLE	2	3	4	5 COMFORTABLE	☐
Telephone privacy:	1 UNCOMFORTABLE	2	3	4	5 COMFORTABLE	☐
Electric lighting **comfort:**	1 UNCOMFORTABLE	2	3	4	5 COMFORTABLE	☐
How bright it gets:	1 TOO BRIGHT	2	3	4	5 COMFORTABLE	☐
Glare from lights:	1 UNCOMFORTABLE	2	3	4	5 NO GLARE	☐

Access to daylight: 1   2   3   4   5   ☐
                                    INADEQUATE   ADEQUATE

OVERALL, WOULD YOU SAY THAT YOUR WORKSPACE HELPS OR HINDERS YOU IN YOUR WORK?
                                              1   2   3   4   5   ☐
                                    MAKES WORK MORE   MAKES WORK
                                       DIFFICULT   EASIER

GENERALLY, HOW SATISFIED ARE YOU WITH THE PHYSICAL ENVIRONMENT IN WHICH YOU WORK?
                                              1   2   3   4   5   ☐
                                    DISSATISFIED   SATISFIED

A.4   NUTAU – Research Center for Architecture and Urban Design Technology

USP – UNIVERSITY OF SÃO PAULO, BRAZIL

POE – POST-OCCUPANCY EVALUATION

- The objective of this research is the refinement and improvement of the quality of office buildings. Your suggestions are essential for this process.
- It is not necessary to identify yourself.

Date: _____   Time of Interview: Start _____   End _____

Sky: ○ clear   ○ partly cloudy   ○ cloudy

Approximate location of Interviewee

Your position on the floor plan below is important in carrying out the thermal comfort analysis.

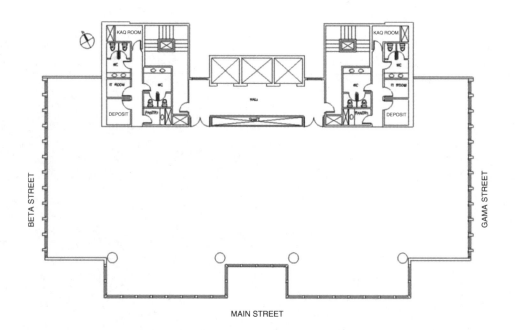

1 – PROFILE OF RESPONDENT

1.1. Gender:
① female   ② male

1.2. Education:
① Primary School Incomplete   ② Primary School Complete   ③ Secondary School Incomplete
④ Secondary School Complete   ⑤ Undergraduate Incomplete   ⑥ Undergraduate Complete

1.3. Family Income: (minimum salaries: R$240,00 per month, as reference)

① 1 to 5   ② >5 to 10   ③ >10 to 20   ④ >20

1.4. Age (years):

① <20   ② 21 to 30   ③ 31 to 40   ④ 41 to 50   ⑤ >50

1.5. Time working in this company (years):

① <1   ② 1 to 5   ③ >5 to 10   ④ >10

1.6. Activity you carry out most frequently in the office:

① reading and/or writing by hand   ② working on computer   ③ drawing/design on paper
④ talking on telephone   ⑤ other (which?) _____

1.7. Frequency of internal and/or external movement for the purposes of work activities:

① low   ② average   ③ high

1.8. On a typical workday, indicate the percentage of time that you spend:

• At your work station   _____
• In a meeting room   _____
• Other places within the office   _____
• Performing work outside the office   _____

_____
100%

1.9. You are a:

① smoker   ② non smoker

1.10. On a typical workday, how many times do you leave your work station to smoke?

① not applicable   ② 1 to 3   ③ >3 to 5   ④ >5

2 – NEIGHBOURHOOD

2.1. Distance from residence to workplace (km):

① <5   ② >5 to 10   ③ >10 to 15   ④ >15 to 20   ⑤ >20

2.2. Most frequent mode of transport to workplace:

① municipal bus   ② inter-municipal bus   ③ your own or company vehicle
④ two or more modes of transport   ⑤ share a ride   ⑥ other (which?) _____

2.3. Average duration of travel from residence to workplace (minutes):

① <30   ② >30 to 45   ③ >45 to 60   ④ >60 to 90   ⑤ >90

2.4. How often do you travel around the neighbourhood of your office building?

① never   ② occasionally   ③ daily

2.4.1. If never, why?

① the area is dirty   ② lack of time   ③ the area is dangerous   ④ other (which?)

2.5. If you do travel around the neighbourhood, how many blocks do you travel?

① 1 to 2   ② 3 to 5   ③ 6 to 10   ④ >10

2.6. You make use of the neighbourhood for (please list activities):

	① Never	② Occasionally	③ Daily	Favourite location (e.g.: shopping mall, business, park)	Most common means of travel (e.g.: walking, car, bus)
2.6.1. Leisure					
2.6.2. Meals					
2.6.3. Shopping					
2.6.4. Services					
2.6.5. Parking					

3 – THE BUILDING

Indicate in the box below your opinion about aspects of the building.

	Not applicable	☹☹ Very Poor	☹ Poor	☺ Good	☺☺ Excellent	If ☹ poor or ☹☹ terrible, why?
3.1. Pedestrian access to the building						
3.2. Access for physically disabled						
3.3. Exit routes in case of emergency						
3.4. Fire safety						
3.5. Security against intruders						
3.6. Waiting time for elevators						
3.7. Elevator safety (functioning)						
3.8. Location of stairways						
3.9. Signage for interior of building						
3.10. External appearance of building						
3.11. Parking						
3.12. Cleanliness of common areas of building						
3.13. Speed and efficiency of maintenance services						
3.14. Water Quality						

4 – WORK ENVIRONMENT

4.1. LAYOUT/FURNITURE

4.1.1. Are there places on your floor that are suitable for users?

① yes   ② no   Which ones? _____

4.1.2. Are there places on your floor that are disagreeable for users?

① yes   ② no   Which ones? _____

4.1.3. Did you participate in designing the layout for your work environment?

① no   ② somewhat   ③ totally

4.1.4. In the table below, indicate your impressions regarding:

	Not applicable	☻☻ Very Poor	☹ Poor	☺ Good	☺☺ Excellent	If ☹ poor or ☻☻ very poor, why?
4.1.4.1. Distance between other areas of activity in which you are involved						
4.1.4.2. Distance between you and your immediate supervisor						
4.1.4.3. Size and arrangement of your workplace in relation to the activities that you carry out						
4.1.4.4. Space available for storage of materials at your workstation						
4.1.4.5. Visual privacy at your workstation						
4.1.4.6. Telephone privacy at your workstation						
4.1.4.7. Height of partition at your workstation						
4.1.4.8. Height of your desk or work area						
4.1.4.9. Type of chair where you sit						
4.1.4.10. Possibility of adjustment of chair						
4.1.4.11. Ease of making these adjustments						
4.1.4.12. Location of meeting rooms						
4.1.4.13. Space for formal meetings on the floor						
4.1.4.14. Space for informal meetings						
4.1.4.15. Location of this space						
4.1.4.16. Space for storage, files, storerooms on your floor						

	Not applicable	☹☹ Very Poor	☹ Poor	☺ Good	☺☺ Excellent	If ☹ poor or ☹☹ very poor, why?
4.1.4.17. Location of storage area						
4.1.4.18. Location of printing area on your floor						
4.1.4.19. Location and characteristics of hallways						
4.1.4.20. Location and characteristics of stairways and elevators						
4.1.4.21. Location of washrooms						
4.1.4.22. Size of washrooms						
4.1.4.23. Speed and efficiency of technical maintenance						
4.1.4.24. Cleanliness on your floor						
4.1.4.25. Fire safety						
4.1.4.26. Security against theft						
4.1.4.27. Distance between you and your work-mates						
4.1.4.28. Distance between you and equipment making uncomfortable noise						
4.1.4.29. Appearance of the layout on the floor						
4.1.4.30. Access and circulation for physically disabled						
4.1.4.31. Quality of drinking water						

4.2. COMFORT

4.2.1. In your work environment, what is your experience of the aspects noted in the following table in **WINTER**?

	Not Applicable	☹☹ Very Poor	☹ Poor	☺ Good	☺☺ Excellent	If ☹ poor or ☹☹ very poor, why?
4.2.1.1. Temperature						
4.2.1.2. Humidity						
4.2.1.3. Air quality						
4.2.1.4. Ventilation						
4.2.1.5. Odours						
4.2.1.6. Lighting quality						

4.2.2. In your work environment, what is your experience of the performance noted in the following table in **SUMMER**?

	Not applicable	☹☹ Terrible	☹ Poor	☺ Good	☺☺ Excellent	If ☹ poor or ☹☹ terrible, why?
4.2.2.1. Temperature						
4.2.2.2. Humidity						
4.2.2.3. Air quality						
4.2.2.4. Ventilation						
4.2.2.5. Odours						
4.2.2.6. Lighting quality						

4.2.3. How would you characterize the glare on your computer screen?
① no glare   ② occur occasionally   ③ occur 50% of the time   ④ occur frequently

4.2.4. Indicate the origin of the glare, if this is a problem for you.
① artificial lighting   ② natural lighting   ③ don't know

4.2.5. Mark in the following table the level of distraction caused by **noise**.

	Not applicable	☹☹ Constant distractions	☹ Frequent distractions	☺ Distractions are rare	☺☺ No distractions
4.2.5.1. Background noise (voices/equipment)					
4.2.5.2. Air conditioning/ ventilation systems					
4.2.5.3. Artificial lighting					

4.3. ENVIRONMENTAL CONTROL

4.3.1. Mark in the following table how you feel about the possibilities for control of the indicated items in your work environment.

	Not applicable	☹☹ Very Unsatisfied	☹ Unsatisfied	☺ Satisfied	☺☺ Very satisfied	If ☹ or ☹☹, why?
4.3.1.1. Temperature						
4.3.1.2. Ventilation						
4.3.1.3. Blinds						
4.3.1.4. Artificial lighting						
4.3.1.5. Degree of freedom to control the above items						

5 – CONCLUSIONS

5.1. In your opinion, which of the items below do you think should be improved in your work environment (put an X in the table below).

5.1.1. Layout	
5.1.2. Work stations	
5.1.3. Meeting rooms	
5.1.4. Storage	
5.1.5. Equipment	
5.1.6. Coffee area	
5.1.7. Washrooms	
5.1.8. Temperature	
5.1.9. Ventilation	
5.1.10. Lighting	
5.1.11. Noise levels	
5.1.12. Privacy	
5.1.13. Aesthetics	
5.1.14. Smoking area	
5.1.15. None	
5.1.16. Other(s)	(which?)

5.2. Do you feel that the environment in which you work is agreeable?
① not at all   ② somewhat   ③ agreeable   ④ very agreeable

5.3. Are you concerned about this?
① not at all   ② somewhat   ③ concerned   ④ very concerned

5.4. In your personal evaluation, classify (from 1 to 10), in order of importance in a work environment, the items indicated (with '1' as the most important).

5.4.1. Thermal comfort	
5.4.2. Air quality	
5.4.3. Visual comfort (lighting/shades)	
5.4.4. Acoustic comfort	
5.4.5. Acoustic and visual privacy	
5.4.6. Comfort of furnishings and dimensions of workstation	
5.4.7. Fire safety	
5.4.8. Security against theft	
5.4.9. Liberty to control the conditions of your own work environment	
5.4.10. Beauty/aesthetics of the building	

5.5. Taking into consideration all of the aspects already analysed above, how do you feel in relation to the office where you work?

① very unsatisfied   ② unsatisfied   ③ satisfied   ④ very satisfied

5.6. In terms of the environmental conditions, how would you characterize this building as a workplace?

① terrible   ② poor   ③ good   ④ excellent

5.7. Make any comments here that you think are necessary

_____

_____

_____

_____

_____

Thank you very much for your valuable cooperation

A.5   Building user survey questionnaire, the Netherlands

Selected questions regarding building and location

The following questions relate to your satisfaction with a number of building aspects at the time of your move and how important those factors are to you. Please provide an explanation beside your answers.

Satisfaction					Remarks	Importance			Explanation
very unsatisfied	unsatisfied	neutral	satisfied	very satisfied		not important	neutral	important	
Building Location									
					Accessibility by bicycle				
					Distance to stores				
The Building									
					Image of the exterior				
					Coffee service (coffee, tea, pop, etc.)				
Your Workspace									
					Amount of space				
					Privacy				
					Daylight				
					Noise				

Excerpt from IBPE-NL toolkit: Occupant Questionnaire (translated into English) used in Rijnland Water Board Case study (Chapter 15).

A.6   Measuring the effects of innovative working environments

The whole questionnaire consists of two parts: A. closed questions, and B. open questions. Both parts are meant to measure people's perception and use of their working environment before and after moving into an innovative office, i.e. transparent and non-territorial. The questionnaire goes into background information (personal characteristics such as sex and age), organizational characteristics, work processes (e.g. % of time people use for different activities), user satisfaction and overall appraisal (e.g. wanting back to the old situation yes or no and why). The list below shows some examples.

Please rate your satisfaction about the following aspects of your working environment?

	very dissatisfied				very satisfied
auditory privacy (not being disturbed by sounds)	1	2	3	4	5
conversation privacy (not being overheard)	1	2	3	4	5
visual privacy (not being seen)	1	2	3	4	5
possibility to influence the office climate (light, sunblinds, ventilation)	1	2	3	4	5
natural light	1	2	3	4	5
view from your workplace	1	2	3	4	5
general office climate	1	2	3	4	5
size of your workplace	1	2	3	4	5
convenience	1	2	3	4	5
image	1	2	3	4	5
flexibility of your working environment	1	2	3	4	5
distance to your colleagues	1	2	3	4	5
sharing of workplaces	1	2	3	4	5
making reservations for your workplace	1	2	3	4	5
ad-hoc using of workplaces	1	2	3	4	5
functionality (suitability) of the workplaces	1	2	3	4	5
adjusting your desk	1	2	3	4	5
adjusting your chair	1	2	3	4	5
moving your trolley	1	2	3	4	5

How would you rate the level of disturbance in your new working environment?

	less disturbed	the same	more disturbed
questions of colleagues	☐	☐	☐
conversations of colleagues with others	☐	☐	☐
telephoning of colleagues	☐	☐	☐
walking around of colleagues	☐	☐	☐

How are the following aspects changed in your new working environment?

	decreased	unchanged	increased
contact with your colleagues	☐	☐	☐
contact with your supervisor(s)	☐	☐	☐
team spirit of your department	☐	☐	☐

Is it in your new working environment *less* or *more* difficult to solve problems?

☐ less difficult   ☐ the same   ☐ more difficult   ☐ other, i.e. ....

How does your new working environment influence your productivity?

☐ negative   ☐ positive   ☐ I do not know   ☐ other: ....

How would you mark your productivity in your new working environment?

1	2	3	4	5	6	7	8	9	10

very
low

very
high

I would rather not return to my old working environment?

☐ disagree   ☐ neutral   ☐ agree   ☐ I do not know

What is your general impression of your new working environment?

☐ negative   ☐ positive   ☐ I do not know

Which 3 aspects in your working environment are **the most positive** to do your work?

1.
2.
3.

Which 3 aspects in your working environment are **the most negative** to do your work?

1.
2.
3.

Source: Vos, P.G.J.C. and G.P.R.M. Dewulf (1999), *Searching for Data. A method to evaluate the effects of working in an innovative office*. Delft University Press. The questionnaire is updated and extended with other methods and modules on health, imago, flexibility, costs and the implementation process. The new method will be published in Volker, L. and D.J.M. van der Voordt (2004).
Werkomgevingsdiagnose-instrument: methoden voor het meten van prestaties van kantoorhuisvesting [a diagnostic tool to measure the performance of physical working environments]. Delft: Center for People and Buildings. In press.

A.7   Three measurement tools from Germany

1. Mapping sentence of work efficiency in office environments (Walden, 1999, 2003)

The occupant, expert or visitor (p) assesses his/her own total impression with reference to

a = areas

the different <u>areas</u> of the
office environment

$\left\{\begin{array}{l} a1 \ = \text{façade/landscape/site} \\ a2 \ = \text{entrance area} \\ a3 \ = \text{restaurant/break area} \\ a4 \ = \text{circulation area} \\ a5 \ = \text{conference rooms} \\ a6 \ = \text{single and executive offices} \\ a7 \ = \text{shared and group offices} \\ a8 \ = \text{air, heat, cooling, noise, sanitary} \\ a9 \ = \text{safety, security} \\ a10 = \text{entire building} \end{array}\right\}$ regarding

c = criteria

different <u>criteria</u>

$\left\{\begin{array}{l} c1 = \text{functional} \\ c2 = \text{aesthetic} \\ c3 = \text{social} - \text{physical} \\ c4 = \text{ecological} \\ c5 = \text{organizational} \\ c6 = \text{economic} \end{array}\right\}$ **with regard to**

the effect on his/her <u>present</u> (p)
work efficiency $\Rightarrow$ as

$\left\{\begin{array}{ll} p1 = \text{very good} & (+2) \\ p2 = \text{good} & (+1) \\ p3 = \text{neither nor} & (+0-) \\ p4 = \text{bad} & (-1) \\ p5 = \text{very bad} & (-2) \end{array}\right\}$ **and on**

<u>importance for the future</u> (i) $\Rightarrow$ as

$\left\{\begin{array}{ll} i1 = \text{very important} & (+2) \\ i2 = \text{important} & (+1) \\ i3 = \text{neither nor} & (+0-) \\ i4 = \text{unimportant} & (-1) \\ i5 = \text{very unimportant} & (-2) \end{array}\right\}$

2. System to judge office quality (Walden, 1999, 2003). Part one

Criteria Office environment	Façade/landscape/site	Entrance area	Restaurant/break area	Circulation areas/corridors/elevators
Functionality (usefulness and time/energy-saving factors)	– Artificial/natural lighting – Handicapped accessibility – Resistant, easy-care material – Noise insulation – Maintenance – Security – Parking – Delivery area – Trash removal	– Artificial/natural lighting – Handicapped accessibility – Elevators – Security barrier – Service area – Acoustics – Flexibility – Noise insulation – For motor vehicles (deliveries and emergencies) – For pedestrians	– Artificial/natural lighting – Handicapped accessibility – Restrooms – Acoustics – Flexibility – Noise insulation	– Artificial/natural lighting – Handicapped accessibility – Orientation – Connection between offices – Automatic light switches – Mail room/service area – Waiting time for elevators
Stylistic – aesthetic (appearance of exterior/interior architecture)	– Transparency – Façade – Green space – Curved forms – Adjustment to arrangement of the surroundings/contextuality – Quality of surroundings – Cleanliness – Vandalism, graffiti	– Spacious, friendly vs. cramped, cold – Plants – Pictures – Works of art – Fountains – Cleanliness	– Plants – Pictures – Works of art – Fountains – View – Cleanliness	– Plants – Pictures – Works of art – Fountains – Cleanliness
Social – physical (value for communication)	– Seating arrangements	– Seating arrangements (flexible) – Opportunities for privacy	– Seating/grouping of chairs and tables/area dividers	– Seating/grouping of chairs – Opportunities for privacy
Ecological value (value of health)	– Energy-saving – Roof gardens – Winter gardens – Exterior facilities – Safety – Quality of materials	– Scent – Adequacy of the rooms/layout – Temperature/ventilation – Quality of materials – Energy-saving – Good isolation – Window space	– Scent/smoke – Adequacy of the rooms/layout – Temperature/ventilation – (Pleasant, bright) materials – (Comfortable) atmosphere – Energy-saving	– (Short, wide, bright) corridors – Natural lighting
Organizational (planning, construction process and later administration)	– Infrastructure, shops, banks – Transport connections – Hotels, restaurants – Libraries, opportunities for leisure time activities	– Orientation – Easily reachable centre – Information, personal reception	– Food distribution – Cashier – Tables – Food quality	– Departments with the same or similar duties close together – Time and energy-saving layout

3. System to judge office quality (Walden, 1999, 2003). Part two

Criteria / Office environment	Conference rooms	Single and executive offices	Shared and group offices	Ventilation/heat/cooling/ noise/restrooms/security	Entire building
Functionality (usefulness and time/energy-saving factors)	– Artificial/natural lighting – Handicapped accessibility – Flexible seating – Acoustics – Flexibility – Noise insulation – Technical equipment such as beamer, overhead projectors, computers	– Artificial/natural lighting – Handicapped accessibility – Personal changes for usage – Individual light control – Orientation – Windows – Layout of the office – Acoustics – Flexibility – Noise insulation – Computer work space – Access to service area	– Artificial/natural lighting – Handicapped accessibility – Personal charges for usage – Individual light control – Orientation – Windows – Layout of the office – Acoustics – Flexibility – Noise insulation – Computer work space – Access to service area	– Self-control of ventilation/ heating/cooling – Ability to automatically close the windows in storms – Cleanliness of the restrooms	– Artificial/ natural lighting – Handicapped accessibility – Personal changes for usage – Acoustics – Flexibility – Noise insulation
Stylistic – aesthetic (appearance of exterior/ interior architecture)	– Plants, pictures, works of art – Cleanliness – Divider walls/doors – View – Materials	– Plants – Pictures – Works of art – Cleanliness – Embellishments, personalization	– Plants – Pictures – Works of art – Cleanliness – Embellishments, personalization	– Comfort of the restrooms	– Plants – Pictures, works of art – Fountains – Cleanliness – Embellishments, personalization
Social – physical (value for communication)	– Open arrangement – Ease of communication – Usage compromise – Comfort telephones – Video conference equipment	– Open arrangement – Ease of communication – Opportunities for privacy (private meetings) – Status demarcation	– Open arrangement – Ease of communication – Opportunities for privacy territoriality – Status demarcation – Team-work	– Noise level of the electronic devices – Surveillance cameras – Alarm system	– Ease of communication – Opportunities for team-work
Ecological value (value for health)	– Scent/smoke – Adequacy of the rooms/layout – Temperature/ventilation – Energy-saving/self- regulating heat – Natural air flow – Air conditioning – Protection from the sun – Ergonomic adjustment of the furniture – Coffee-making facilities	– Scent – Adequacy of the rooms/layout – Temperature/ventilation: energy-saving/self-regulating – Personal changes – Environmental control in regulation of the ambient environment – Natural air flow, openable windows – Air-conditioning – Protection from the sun – Ergonomic adjustment – Coffee-making facilities	– Scent/smoke – Adequacy of the rooms/layout – Temperature/ventilation – Energy-saving/self- regulating heat – Personal changes – Opportunities for environmental control – Natural air flow – Air-conditioning – Protection from the sun – Ergonomic adjustment – Coffee-making facilities	– Individual control of ventilation – Individual control of heating/cooling – Individual control of air-conditioning – Modern, energy-saving central heating – Protection from the sun – Hygiene of the restrooms – Safety – Fire extinguishers, exits – Sprinkler system – Smoke alarms	– Scent – Adequacy of the rooms/layout – Temperature/ ventilation – Energy-saving/ self-regulating heat – Natural air flow – Air-conditioning – Protection from the sun
Organizational	– Variable room dividers – Technical equipment, access to information	– Clear divisions/self- containment of the office – Access to information	– Clear divisions/self- containment/enclosure of the office	– Fire escape routes	– Connections between rooms/sections

4. IBPE questionnaire

University of Koblenz,
Institute of Psychology
Project Seminar:
Methods of Empirical Social Research
Principal Investigator: Dr. Rotraut Walden
Team: Names & Phone
in Cooperation with
Prof. Dr. Wolfgang F.E. Preiser©[1]

Last name: _____   First name: _____
Age: _____   Sex: ☐ female   ☐ male
Phone: _____   Profession: _____

Performance evaluation

The purpose of this evaluation is to assess how well this building performs. The first aim of this building profile is to evaluate this building in terms of work efficiency and the well-being of the users at the moment (1). Secondly, we would like to know how important the aspects of this building are for the future (2). The benefits of the evaluation include: identification of good and bad performance aspects of the building, better building utilization, feedback on how to improve similar buildings in the future, and remodelling of your own building.

We want to thank you in advance for your efforts. We assure you all data will be kept confidential. Your name is only necessary for further inquiry in case we have further questions about your answers. If there is a question which is not applicable to you, you can skip that question by marking the appropriate blanks with an 'X' at the column 'n.a.' (not applicable).

The key for quality ratings is:

Evaluation of the building concerning the work efficiency at the moment	Very good	Good	Neutral	Bad	Very bad	Not applicable
	☺☺	☺	☹	☹	☹☹	
	+2	+1	+0−	−1	−2	n.a.
Evaluation of the importance of the aspects for work efficiency in future	Very important	Important	Neither important, nor unimportant	Unimportant	Very unimportant	Not applicable

In the survey that follows, please respond to all the questions.
Thank you very much for your cooperation!

1   Adaptation to an user-needs analysis by Rotraut Walden

1. Evaluation of the overall quality of the building

	(1) Evaluation of the building concerning the work efficiency at the moment						(2) Evaluation of the importance of the aspects for work efficiency in future					
	+2 ☺☺	+1 ☺	+0− ☺	−1 ☹	−2 ☹☹	n.a.	+2 ☺☺	+1 ☺	+0− ☺	−1 ☹	−2 ☹☹	n.a.
a) Aesthetic quality of exterior												
b) Aesthetic quality of interior												
c) Amount of space												
d) Environmental quality (lighting, acoustics, temperature, etc.)												
e) Proximity of views												
f) Adaptability to changing uses												
g) Security												
h) Handicapped accessibility												
i) Maintenance												
j) Relationship of spaces/layout												
k) Quality of building materials												
(1) floors												
(2) walls												
(3) ceilings												
l) Other: _____												
m) Sound insulation												

2. Evaluation of the overall quality of the building site

	(1) Evaluation of the building concerning the work efficiency at the moment						(2) Evaluation of the importance of the aspects for work efficiency in future					
	+2 ☺☺	+1 ☺	+0− ☺	−1 ☹	−2 ☹☹	n.a.	+2 ☺☺	+1 ☺	+0− ☺	−1 ☹	−2 ☹☹	n.a.
a) Vehicular access												
b) Parking												
c) Delivery												
d) Waste removal												
e) Aesthetic quality of views												
f) Landscaping												
g) Pedestrian access												
h) Other: _____												

3. Space categories

Space A: Entrance Hall
Space B: Restaurants and gardens
Space C: Conference rooms
Space D: Individual offices
Space E: Group offices

A.8   The principles of universal design

Version 2.0 – 4/1/97
THE CENTER FOR UNIVERSAL DESIGN
N.C. State University

Compiled by advocates of universal design, listed in alphabetical order:
Bettye Rose Connell, Mike Jones, Ron Mace, Jim Mueller, Abir Mullick, Elaine Ostroff, Jon Sanford, Ed Steinfeld, Molly Story and Gregg Vanderheiden

Universal design

The design of products and environments to be usable by all people, to the greatest extent possible, without adaptation or specialized design.

Principle one: equitable use

The design is useful and marketable to people with diverse abilities.

Guidelines:

1a. Provide the same means of use for all users: identical whenever possible; equivalent when not.
1b. Avoid segregating or stigmatizing any user.
1c. Make provisions for privacy, security, and safety equally available to all users.
1d. Make the design appealing to all users.

Principle two: flexibility in use

The design accommodates a wide range of individual preferences and abilities.

Guidelines:

2a. Provide choice in methods of use.
2b. Accommodate right- or left-handed access and use.
2c. Facilitate the user's accuracy and precision.
2d. Provide adaptability to the user's pace.

Principle three: simple and intuitive use

Use of the design is easy to understand, regardless of the user's experience, knowledge, language skills, or current concentration level.

Guidelines:

3a. Eliminate necessary complexity.
3b. Be consistent with user expectations and intuition.
3c. Accommodate a wide range of literacy and language skills.
3d. Arrange information consistent with its importance.
3e. Provide effective prompting and feedback during and after task completion.

Principle four: perceptible information

The design communicates necessary information effectively to the user, regardless of ambient conditions or the user's sensory abilities.

Guidelines:

4a. Use different modes (pictorial, tactile) for redundant presentation of essential information.
4b. Maximize 'legibility' of essential information.
4c. Differentiate elements in ways that can be described (i.e. make it easy to give instructions or directions).
4d. Provide compatibility with a variety of techniques or devices used by people with sensory limitations.

Principle five: tolerance for error

The design minimizes hazards and the adverse consequences of accidental or unintended actions.

Guidelines:

5a. Arrange elements to minimize hazards and errors; most used elements, most accessible; hazardous elements eliminated, isolated, or shielded.
5b. Provide warnings of hazards and errors.
5c. Provide fail safe features.
5d. Discourage unconscious action in tasks that require vigilance.

Principle six: low physical effort

The design can be used efficiently and comfortably and with a minimum of fatigue.

Guidelines:

6a. Allow user to maintain a neutral body position.
6b. Use reasonable operating forces.
6c. Minimize repetitive actions.
6d. Minimize sustained physical effort.

Principle seven: size and space for approach and use

Appropriate size and space is provided for approach, reach, manipulation, and use regardless of user's body size, posture, or mobility.

Guidelines:

7a. Provide a clear line of sight to important elements for any seated or standing user.
7b. Make reach to all components comfortable for any seated or standing user.
7c. Accommodate variations in hand and grip size.
7d. Provide adequate space for the use of assistive devices or personal assistance.

Copyright 1997 N.C. State University
Major funding provided by the National Institute on Disability and
Rehabilitation Research

A.9   Information sources for building commissioning

- Building Commissioning Association
  P.O. Box 2016
  Edmonds, Washington 98020
  www.bcxa.org

- Portland Energy Conservation Inc. (PECI)
  1400 SW 5th Avenue, Suite 700
  Portland, Oregon 97201
  www.peci.org

- Architectural Energy Corporation
  2540 Frontier Avenue, Suite 201
  Boulder, Colorado 80301
  www.archenergy.com
  - ENFORMA® Commissioning Toolkit
  - MicroDataLogger™ Portable Data Acquisition System

- Energy Design Resources
  www.energydesignresources.com

- National Environmental Balancing Bureau
  8575 Grovemont Circle
  Gaithersburg, Maryland 20877
  www.nebb.org

Index